Joseph Abruscato   Joan Wade Fossaceca   Jack Hassard   Donald Peck

HOLT SCIENCE

Holt, Rinehart and Winston, Publishers
New York · Toronto · Mexico City · London · Sydney · Tokyo

THE AUTHORS

Joseph Abruscato
Associate Dean
College of Education and Social Services
University of Vermont
Burlington, Vermont

Joan Wade Fossaceca
Teacher
Pointview Elementary School
Westerville City Schools
Westerville, Ohio

Jack Hassard
Professor
College of Education
Georgia State University
Atlanta, Georgia

Donald Peck
Supervisor of Science
Woodbridge Township School District
Woodbridge, New Jersey

Cover photos, front: Harold Sund/The Image Bank; back: Dennis Dilaura/After Image.
The photos on the front and back covers are of Monument Valley in Arizona. This
desert habitat clearly shows the effects of physical weathering and wind erosion.

Photo and art credits on pages 342–343

ISBN: 0-03-003079-X
    89 071 76

ACKNOWLEDGMENTS

Teacher Consultants

Armand Alvarez
District Science Curriculum Specialist
San Antonio Independent School District
San Antonio, Texas

Sister de Montfort Babb, I.H.M.
Earth Science Teacher
Maria Regina High School
Uniondale, New York
Instructor
Hofstra University
Hempstead, New York

Ernest Bibby
Science Consultant
Granville County Board of Education
Oxford, North Carolina

Linda C. Cardwell
Teacher
Dickinson Elementary School
Grand Prairie, Texas

Betty Eagle
Teacher
Englewood Cliffs Upper School
Englewood Cliffs, New Jersey

James A. Harris
Principal
Rothschild Elementary School
Rothschild, Wisconsin

Rachel P. Keziah
Instructional Supervisor
New Hanover County Schools
Wilmington, North Carolina

J. Peter O'Neil
Science Teacher
Waunakee Junior High School
Waunakee, Wisconsin

Raymond E. Sanders, Jr.
Assistant Science Supervisor
Calcasieu Parish Schools
Lake Charles, Louisiana

Content Consultants

John B. Jenkins
Professor of Biology
Swarthmore College
Swarthmore, Pennsylvania

Mark M. Payne, O.S.B.
Physics Teacher
St. Benedict's Preparatory School
Newark, New Jersey

Robert W. Ridky, Ph.D.
Professor of Geology
University of Maryland
College Park, Maryland

Safety Consultant

Franklin D. Kizer
Executive Secretary
Council of State Science Supervisors, Inc.
Lancaster, Virginia

Readability Consultant

Jane Kita Cooke
Assistant Professor of Education
College of New Rochelle
New Rochelle, New York

Curriculum Consultant

Lowell J. Bethel
Associate Professor, Science Education
Director, Office of Student Field Experiences
The University of Texas at Austin
Austin, Texas

Special Education Consultant

Joan Baltman
Special Education Program Coordinator
P.S. 188 Elementary School
Bronx, New York

CONTENTS

SKILLS OF SCIENCE

Frank and Debbie had planned to go sailing on the lake. Then they saw dark clouds in the east. They checked to see from what direction the wind was blowing. The wind was also coming from the east. They called to get a weather report. As they thought, a storm was moving toward them. They did not take the boat out. They tied it up and went home.

Why did Frank and Debbie think a storm was coming? Why did they call for a weather report? Frank and Debbie had **observed,** or noticed, two things. First, they observed dark clouds in the east. Next, they observed that the wind was also coming from the east. They compared these **observations** with what they knew. They knew that dark clouds often bring bad weather. But they also knew that this was not always the case. They needed more information. So they called for a weather report.

Frank and Debbie did some things weather scientists do. A scientist must observe things closely. A scientist can learn a lot about the world through observations. Sometimes a scientist has to check observations with other scientists. The observations of one person may not be enough to give a true picture of the world. The observations of many people may be needed. This is true of weather forecasting.

Observe: To notice something by seeing or using any of the five senses, such as hearing.

Observations: Anything that we can learn by using our senses, such as sight or hearing.

Record: To put down in writing.

Data: Facts, or pieces of information.

Predict: To forecast on the basis of observations or past experience.

There are over 9,000 weather stations around the world. Scientists at each station make observations. They also **record** information about the weather. They then share this information with each other. Pieces of information are called **data.** Scientists use data to forecast, or **predict,** the weather.

Scientists often must make **measurements** to help them make predictions or forecasts. Suppose a storm is moving across the country. Scientists need to know how fast it is moving. They must first measure how far it has moved. Then they measure how long it took to move that far.

Measurements: Observations that are made by counting something. Often an instrument, such as a ruler, is used to make measurements.

Skill Building Activity

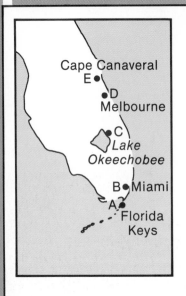

Cape Canaveral
E
D
Melbourne
C
Lake
Okeechobee
B Miami
A
Florida
Keys

You can measure the movement of a storm.

A. Gather these materials: 10 cm (4 in.) thread, piece of cardboard, 5 straight pins, tape, metric ruler, paper, and pencil.

B. Trace the outline map of Florida. Tape your outline to a piece of cardboard.

C. Place a straight pin at each spot where the storm is shown on the map.

D. Tie 1 end of the thread around the first pin. Tie the thread to the second pin. Join all 5 pins with thread.

E. Use the ruler to measure the length of thread between each pin. Record your results in a table like the one shown.

F. Let 1 cm = 100 km (62 mi). Change your numbers to kilometers.

   1. It took 5 hr for the storm to move from A to B. How fast was it moving?

   2. At the same speed, how long would the storm take to move from C to D?

Positions	Distance (1 cm = 100 km)	
	cm	km
A–B		
B–C		
C–D		
D–E		

The storm that upset Frank and Debbie's sailing plans was a hurricane. A hurricane is a storm with strong winds and heavy rain. About ten hurricanes a year form over the warm waters of the Caribbean Sea. Some of these storms may hit the United States. A hurricane that hits land can cause great damage. Luckily, most hurricanes never hit land.

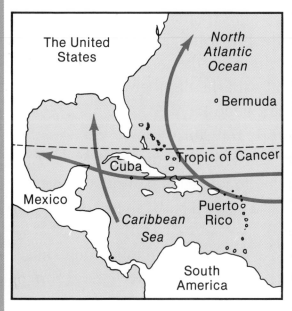

Weather scientists observe hurricanes closely. As a hurricane forms, scientists record information about it on maps. They keep track of its path. Then they can sometimes predict where it will go next. The map on this page shows three paths that a hurricane might follow.

Records are kept in all fields of science. In weather forecasts, good records can save lives.

SKILL BUILDING ACTIVITY

Position	Day
F13	1
F12	2
F11	3
F10	4
F9	5
F8	6

You can track a hurricane and predict where it will go.

A. Gather these materials: paper, pencil, and metric ruler.

B. Trace the map shown on page 7 on a piece of paper. Use a ruler to help you draw straight lines.

C. Look at the hurricane table. Mark where the hurricane is each day on your map. Label each point with the number of the day. (Day 1 has been done for you.)

    1. Where is the storm on day 2?

    2. What city is it near on day 6?

D. Draw a line to join each point. Compare the storm's path to the 3 paths shown on the map on page 5.

    3. Suppose the hurricane follows one of the 3 paths. Predict where it will be on day 7.

    4. Pretend you are a weather forecaster. What cities and countries would you warn about the hurricane?

E. Use the metric ruler and the scale of kilometers to answer question 5.

    5. How far is Miami from San Juan? Suppose the storm turns toward Miami. It is moving at 20 kilometers per hour (kph). How long will it take to hit the city?

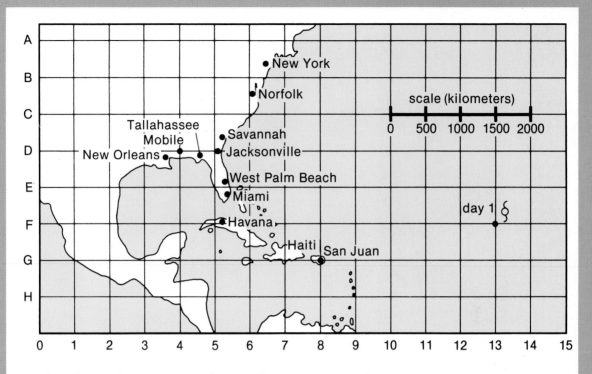

The chart shows some things that scientists do. How did Frank and Debbie act like scientists?

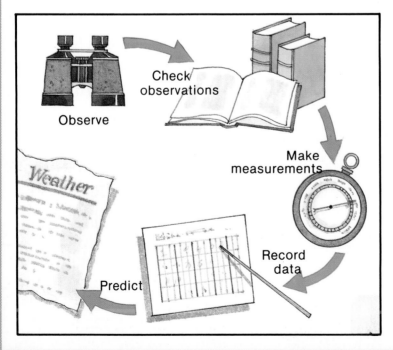

Observe

Check observations

Make measurements

Record data

Predict

Weather

THE CHANGING EARTH

UNIT 1

CHAPTER 1

THE WHOLE EARTH

1-1.

Inside the Earth

Pretend that you are in the machine that is shown in the picture above. The machine was made for use in a movie. In this movie, people took a trip to the center of the earth. The machine is able to drill straight through hard rock.

Imagine drilling to the center of the earth. What would you find? When you finish this section, you should be able to:

☐ **A.** Name the layers of the earth.

☐ **B.** Compare the layers of the earth to the inside of an apple.

☐ **C.** Describe how the three layers of the earth are different from each other.

How would you find out what the inside of a softball is like? It would be simple. Just rip the cover off the ball. Then unwind the string like the boy in the picture.

Finding out what the inside of the earth is like is not as easy. We cannot peel the earth open to look inside. We don't have machines that can travel deep into the earth. But clues from earthquakes and volcanoes help us make guesses about the inside of the earth. You will know more about these clues when you finish this unit.

Scientists who study the earth are called **geologists** (jee-**ahl**-oh-jists). *Geologists* think the earth is made of three layers. You can compare the layers of the earth to the inside of an apple. Look at the pictures on page 12. Both the apple and the earth have three layers. What are the three layers of the apple? What are the three layers of the earth?

Imagine we can go to the center of the earth. We begin with the top layer of the earth. It is mostly rock. It is called the **crust**.

Geologist: A scientist who studies rocks and other features of the earth's layers.

Crust: The thin, solid, outer layer of the earth.

11

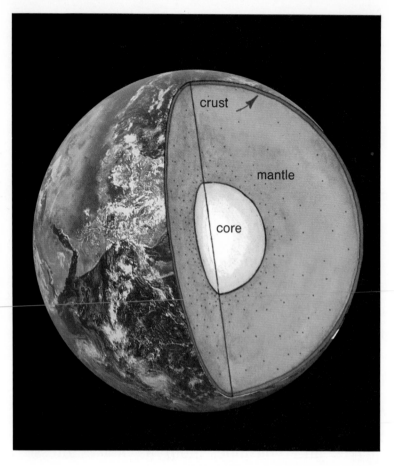

The *crust* is very important to us. It is where we live. It is like a giant treasure chest. Oil, coal, gas, metals, rocks, water, and plants are some of its treasures.

Pretend the earth is the size of an apple. The crust will be as thin as the apple's skin. The earth's crust is very thin in some places. It is thicker in others. It is from 5 to 55 kilometers (3–34 miles) thick.

Let's go deeper into the earth. Below the crust is a layer called the **mantle** (**man**-tuhl). It is about 2,900 kilometers (1,800 miles) thick. The *mantle* is made of mostly solid rock. If the earth

Mantle: A thick, hot layer found under the crust of the earth.

were the size of an apple, the mantle would be as thick as the white part under the skin. Geologists believe the mantle is as hot as 3,000° Celsius (5,400° Fahrenheit). The mantle is hot enough to melt rock.

As we continue our trip into the earth, we come to the third layer. This layer is called the **core**. The *core* of the earth is in the same position as the core of an apple. It is about 3,500 kilometers (2,200 miles) thick. The temperature at the center of the earth is very hot. It may be as high as 4,000° Celsius (7,200° Fahrenheit). This is hot enough to melt the metals. Scientists think that the core has two parts. The outer core is melted iron and nickel. The inner core is solid iron and nickel. If you could see the core, it probably would be glowing hot.

Core: The center of the earth; it is made of nickel and iron.

ACTIVITY

How is an apple like the earth?

A. Gather these materials: an apple and a dinner knife.

B. Look at the picture. Cut the apple in half. Draw a picture of the inside of the apple. Label the crust, mantle, and core.

C. Do not taste things used in a science activity. Your teacher will tell you what to do with the apple.

People did not always know about the earth's layers. In 1909, Andrija Mohorovicic (Ahn-**dree**-ha Mo-ho-**ro**-veh-chick) first found that the earth has layers. The boundary between the crust and mantle was named for him. It is called the Moho. Scientists had hoped to drill a hole through the crust to reach the mantle. They called this hole the Mohole.

Section Review

Main Ideas: The following chart shows the main ideas in this section.

Layer of Earth	Thickness	Made of	Temperature
Crust	5 to 55 km	solid rock	cool
Mantle	2,900 km	mostly solid rock	hot, 3,000°C
Core	3,500 km	nickel, iron solid inner, liquid outer	very hot, 4,000°C

Questions: Answer in complete sentences.

1. What are the three layers of earth? How are they different from each other?
2. Which layer of the earth (a) has the hottest rocks? (b) would take you the longest to travel through? (c) are living things found on?
3. What kinds of problems would the pilot in the picture on page 10 have had if he had really drilled through the earth?

Did you know that over 1 million earthquakes occur each year? Most of them cause little damage. But now and then a strong earthquake hits. If it happens in a city, the damage to buildings could be great. Many people could die.

Look at the picture. It was taken after an earthquake hit Anchorage, Alaska. What damage did this earthquake cause? When you finish this section, you should be able to:

☐ **A.** Describe the changes in the earth's crust during an *earthquake*.

☐ **B.** Describe two ways that geologists measure *earthquakes*.

☐ **C.** Describe the damage that *earthquakes* can cause.

Sudden movements in the earth's crust are called **earthquakes** (**erth**-kwakes). During an *earthquake*, the earth's crust shakes. An earthquake occurs when the crust moves, buckles, cracks, or bends. The earth's crust is very thin compared with the rest of the earth. It is thin like the skin of an apple. The crust of the earth can wrinkle like the skin of an apple.

Most earthquakes are not deep in the earth. They are within 60 km (36 miles) of the surface of the crust. A few earthquakes are deeper in the earth. Some are as deep as 650 km (400 miles).

Geologists keep records of where earthquakes happen. Look at the map on this page. Each red dot shows where a large earthquake has taken place in recent years. Earthquakes occur in some parts of the crust more than in others. These red dots make up what is called the "earthquake belt." Earthquakes happen here more often than in other places on earth.

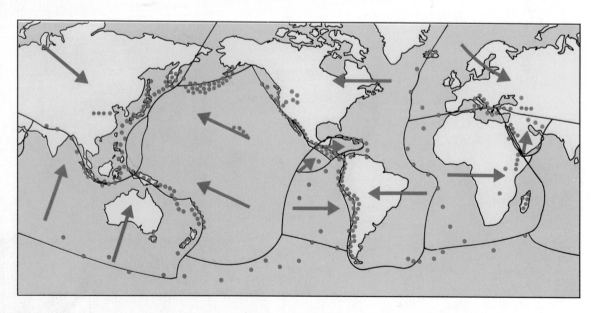

An earthquake causes objects on the surface of the crust to shake. Loose soil and rocks may slide down hillsides. Buildings made of stone or clay may crack. There is a chance of fire when buildings shake. Gas pipes can snap. The gases may catch fire when they leak out of the pipes. In 1906, the city of San Francisco had a big earthquake. After the quake, fires broke out. Almost all of the houses were made of wood. Most of the city burned to the ground.

Rocks in the crust can crack and move. One place where rocks move is called a **fault**. *Faults* are cracks in the earth's crust. Rocks along the sides of a fault may be stuck together. These

Fault: A crack in the earth's crust where rocks can move.

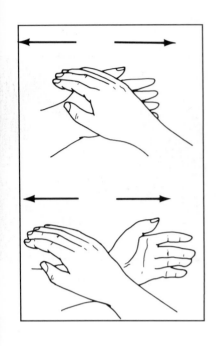

rocks can snap apart. Sometimes the rocks just slip past one another. Press the palms of your hands together. Look at the picture. Your hands will suddenly slide over one another. If this happened to rocks, an earthquake would occur.

Geologists have found ways to measure earthquakes. In 1902 an Italian scientist named Mercalli invented a scale. The scale is based on descriptions of the damage caused by earthquakes. A type I earthquake cannot be noticed by our senses. A type XII earthquake is the strongest. It causes total destruction.

Today scientists have a much more accurate way to measure earthquakes. This method was invented in 1927 by an American scientist named Richter. A machine called a seismograph is used. A picture of one is shown here. The ink pen is attached to a long arm. The shaking of an

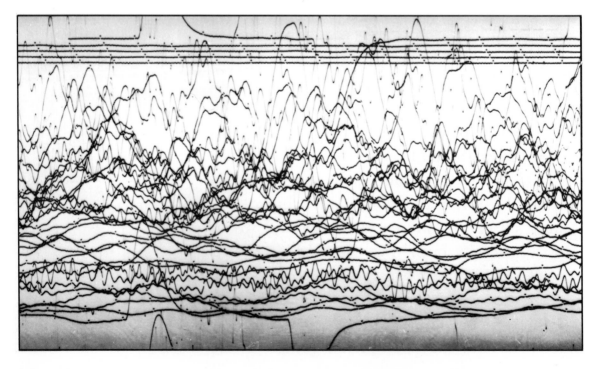

earthquake causes the pen to move up and down. The stronger the earthquake, the farther the pen moves. The moving pen marks the paper on the turning drum. These markings are the records of the crust's movement. The picture on page 18 is a seismograph record of an earthquake. Geologists can measure the height of the pen marks to find out how strong the earthquake was. A number from 1 to 9 is then assigned to the earthquake. An earthquake that is 7 on the Richter scale is called a major earthquake. Any earthquake measuring 8 or 9 causes total destruction.

Section Review

Main Ideas: Earthquakes are sudden movements in the crust. They are caused by the movement of rocks along cracks, or faults. Earthquakes can be mild or strong. Strong ones cause landslides and damage to buildings. Earthquakes are measured with a seismograph and are assigned a number on the Richter scale.

Questions: Answer in complete sentences.

1. What are two ways that an earthquake can be measured?

2. Look at the map shown here. Would an earthquake be most likely at point A, B, or C? How do you know?

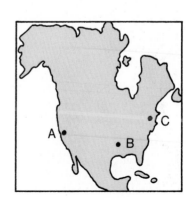

1-3.

Volcanoes

Volcano: An opening in the earth's crust, through which lava escapes.

For hundreds of years, Mt. St. Helens was a quiet snow-capped mountain. Then geologists began to see strange things near the mountain. Earthquakes began to increase in number. The ground around the mountain began to swell. Soon, geologists were saying that Mt. St. Helens was going to blow its top. And it did. The earth rumbled, and the ground split open.

Campers were trapped in the forests around Mt. St. Helens. A huge blast of hot gases and ash spread out over the mountain. Mt. St. Helens was now an active **volcano** (vohl-**kay**-no). What caused the *volcano* to erupt? When you finish this section, you should be able to:

☐ **A.** Describe the stages in which a *volcano* forms.

☐ **B.** Explain how *magma* and *lava* are alike and how they are different.

Some rocks deep under the earth's surface are so hot that they melt. This melted rock is called **magma** (**mag**-ma).

Magma forms in the earth's mantle. It rises toward the surface because it is lighter than solid rock. The magma may push its way out through a crack or weak spot in the earth's crust. If magma reaches the surface, it is called **lava** (**lah**-vah). Steam, rocks, and hot gases may be mixed with *lava*. This hot material piles up, cools, and hardens.

Magma: Red-hot melted rock under the earth's crust.

Lava: Red-hot melted rock coming out of the earth's crust.

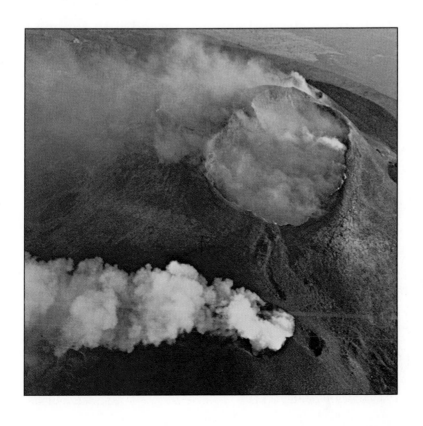

The diagrams below show how a volcano forms. (1) Magma forms in a pool deep in the earth. (2) Cracks form in the rocks above the magma pool. The magma slowly moves up the cracks. (3) As the magma nears the surface, pressure builds up. The rocks begin to push with great force. Sometimes the magma oozes out. At other times the volcano explodes. Material is thrown into the air. Magma also can reach the surface and flow out on the land. Geologists call this a lava flow. (4) A mountain formed in this way is called a volcanic mountain.

The materials that come out of volcanoes are clues. They tell us about the inside of the earth. The inside of the earth is very hot. The inside of the earth is always changing.

Mt. St. Helens is an exploding volcano. When it erupts, rocks and ash are tossed into the air.

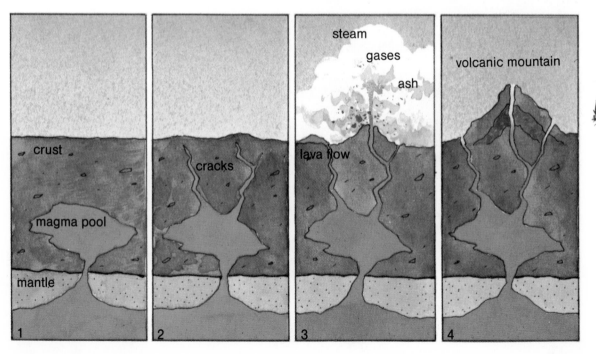

The ash and rock cool as they fall. They pile up around the volcano opening. Finally, a cone-shaped mountain forms.

Not all volcanoes explode like Mt. St. Helens. All the Hawaiian islands are the tops of volcanoes. They reach up from the ocean floor. There are two active volcanoes on Hawaii Island. The one in the photo on this page is called Mauna Loa (Ma-oo-na **Lo**-a). Mauna Loa erupts quietly. The lava is very hot. It flows down the sides of the mountain like hot tar. The lava burns up the trees and plants in its path. As the lava cools, it slows down and stops.

Look back at the map on page 16. Each blue dot shows an active volcano on the earth's surface. Active volcanoes are erupting or have recently erupted. They form a belt called the "ring of fire." Look at the map. The ring of fire and the earthquake belt are almost in the same place. This makes geologists think that volcanoes and earthquakes may be caused by the same thing.

ACTIVITY

What do rocks from a volcano look like?

A. Gather these materials: a set of volcanic rocks labeled A through E, a glass, and water.

B. Record your observations of each rock. Observe its color and its shape. Look carefully at the surface of each rock. Is it smooth? Does it have holes in it? Is it rough? What happens if you put each rock in a glass of water?

C. Go to a library to find books about geology and rocks. Find out the names of the rocks you observed.

Section Review

Main Ideas: Magma can move through cracks inside the earth. A volcano is a hole in the crust through which the magma escapes. Volcanoes and earthquakes are found in almost the same parts of the crust.

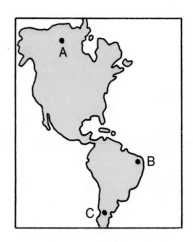

Questions: Answer in complete sentences.

1. What is the difference between magma and lava?
2. Where on the map shown here would you expect to find a volcano?
3. Why do geologists think that earthquakes and volcanoes may have the same cause?

People in Science

Alfred Wegener

Alfred Wegener was a German scientist. In 1912, he wrote a book about the continents. Wegener believed that the continents moved. He could see that the coasts of South America and Africa fit together like the pieces of a puzzle. At one time, Wegener believed, all the continents were part of one super-continent. They slowly drifted apart to form the continents as they are today. Wegener was not widely believed in his own time. But now scientists believe that the continents are still moving apart. This theory is known as Wegener's theory of continental drift.

Wegener went on four expeditions to the North Pole to test his theories. He died in Greenland on his last trip in 1930.

CHAPTER REVIEW

Science Words: Think of a word for each blank. List the letters **a** through **i** on paper. Write the word next to each letter.

People who study rocks and other features of the earth are called ___**a**___. Some of these people study the thin, solid, outer layer of the earth known as the ___**b**___. Some scientists have become interested in the ___**c**___, a thick layer of the earth below the surface layer. Scientists believe there is a hot layer made of metals deep in the earth. It is called the ___**d**___.

Sudden movements of the earth are called ___**e**___. They occur when rocks move along cracks called ___**f**___.

Hot liquid in the crust is called ___**g**___. When it comes to the surface, it is called ___**h**___. The opening in the crust through which the material comes is called a ___**i**___.

Questions: Answer in complete sentences.

1. Make diagrams of an apple and the earth. Draw lines between the parts that are alike. Label the parts.
2. What kinds of damage can an earthquake cause?
3. List the correct order for the eruption of a volcano. (a) Cracks form in the crust. (b) Lava flows. (c) A magma pool forms. (d) Lava cools to form solid rock.
4. Can you predict where earthquakes may occur from this map of volcanoes?

26

CHAPTER 2

THE CRUST MOVES

Imagine going from the United States to France without crossing an ocean. You might only have to walk a few feet. Is it possible? What do you think? When you finish this section, you should be able to:

2-1.

Moving Continents

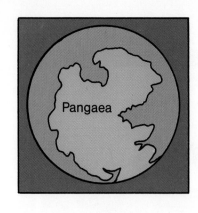

Pangaea

☐ **A.** Explain what is meant by *continental drift*.

☐ **B.** Describe two clues that show that the continents moved.

☐ **C.** Describe or show on a map how scientists think the continents moved.

In 1912, Alfred Wegener wrote that all of the continents were once one piece of land. The diagram shows what all the continents joined might look like. This super-continent was called Pangaea (Pan-**gee**-a). The name Pangaea comes from two Greek words meaning all (pan) lands (gea). Wegener said that Pangaea broke up. The continents moved apart. Geologists call this idea **continental drift**. They believe that the continents are still drifting.

Continental drift: The idea that the continents are moving.

ACTIVITY

Do the continents fit together?

A. Gather these materials: 1 sheet of construction paper, 1 sheet of tracing paper, scissors, paste, and map of the world.
B. Place the tracing paper on top of the map.
C. Trace the outline of North America.
D. Repeat step C for the other continents.
E. Cut out all the tracings.
F. Put the cutouts on the piece of construction paper. Paste the continents together.
  1. Do you think that Pangaea might have existed in the past? Why?

Geologists have found clues that the continents moved. The first clue is the shape of the continents. Look at the maps below. The continents seem to fit like a puzzle. If you could cut them out and move them, they would fit nicely.

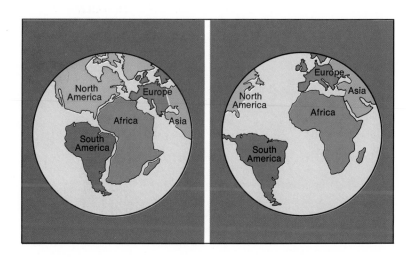

A second clue comes from rocks. The geologists compared the rocks on different continents. Some rocks have traces of plants and animals that lived long ago. These traces in the rock are called **fossils** (**foss**-sills). The picture below shows a *fossil* of a reptile called Mesosaurus (me-zo-**sor**-us). This fossil has been found in only two places. One place is part of South America. The other place is the part of Africa that is right across from it. Mesosaurus lived in lakes and streams. But it was not a good swimmer. It could not travel very far. How could it get across the ocean? Geologists think that when Mesosaurus was alive, South America and Africa were joined. Thus, Mesosaurus didn't have to swim very far.

Fossils: Traces of plants and animals in rocks.

Other clues seem to fit the idea of *continental drift*. Animal fossils have been found under the ice in Antarctica. These animals lived in warm climates. Can the idea of continental drift explain this? Rock colors and layers look alike on continents thousands of miles apart. These clues add up to the idea that the continents were once attached.

Section Review

Main Ideas: The chart describes the clues that support the idea of continental drift.

Clue	How it supports the theory of continental drift
Shape of the continents	The outlines of the continents fit together just like a jigsaw puzzle.
Fossils	The animal could not cross the ocean. Perhaps the continents were joined together.

Questions: Answer in complete sentences.

1. What was Pangaea? What happened to it?
2. What are two clues that the continents may have drifted?
3. Was it ever possible to travel from the United States to France without going over water? Why?
4. Suppose the continents keep drifting. What may happen to the distance between Africa and South America?

This is a giant fault in the crust of the earth. It is along the coast of California. It is called the San Andreas (Sahn Ahn-**dray**-us) fault. The San Andreas fault divides two pieces of the earth's crust. Scientists think these two pieces are rubbing and bumping into each other. What could happen when pieces of the crust rub and bump into each other? When you finish this section, you should be able to:

☐ **A.** Explain that the earth's crust is broken into pieces.

☐ **B.** Describe the kinds of movement that occur where pieces of the crust meet.

☐ **C.** Explain that the movement of pieces of the crust may cause earthquakes and volcanoes.

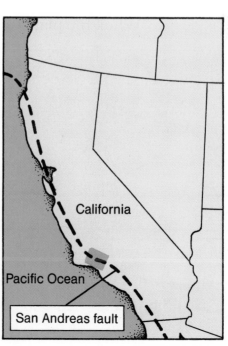

California

Pacific Ocean

San Andreas fault

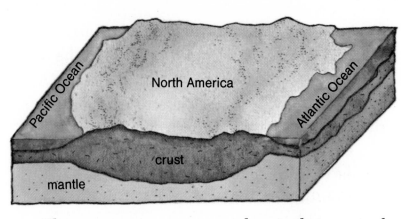

The crust may not cover the earth in a single piece. How is this different from the skin of an apple? Scientists think that the crust is broken into about ten pieces. These pieces are called **plates**. The map shows the major *plates* of the earth's crust. The diagram shows what scientists think each plate looks like.

Plates: Large sections of the earth's crust.

In the last section, you learned that the continents drifted apart. Scientists believe that the continents could drift apart because they were parts of plates. The plates of the crust move. When the plates move, the continents move, too.

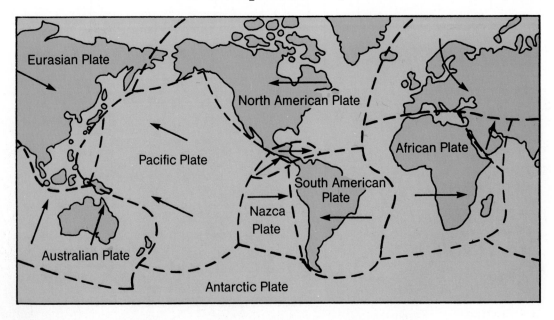

ACTIVITY

How is the earth like an egg?

A. Gather these materials: hard-boiled egg, dinner knife, and paper towel.

B. Keep the egg on the paper towel. Gently tap the egg to crack the shell in several places. Do not remove the pieces.

    1. What layer of the earth is the shell of an egg like?

    2. Why is a broken shell a better model than an unbroken one?

C. Gently push the pieces of shell against each other.

    3. What happens to the shell pieces where they meet?

    4. What is this like on the real earth?

D. Cut through the center of the egg.

    5. What layer of the earth is the yolk like?

    6. What layer is the white part like?

E. Students should not eat materials used in a science activity. Your teacher will tell you what to do with the egg.

Scientists think that plates can move in three ways: (1) The plates can push against each other (collide); (2) the plates can move apart (spread); and (3) they can also slide past each other (slip). Let's look at the three movements. What happens to the earth's crust in each case?

Two plates can push against each other. When this happens, the plates collide. The thin part of

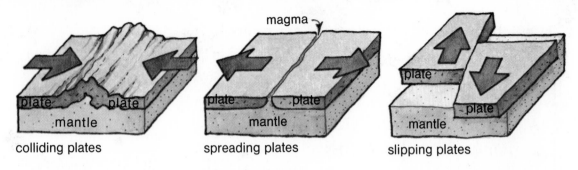

colliding plates   spreading plates   slipping plates

one plate slowly moves under the thick part of the other plate. Earthquakes are very common where plates collide.

Two plates can move apart. As this happens, magma squeezes up between the plates. The magma cools and forms new crust. Volcanoes and earthquakes are common where plates are moving apart.

Two plates can slide past each other. The San Andreas fault divides two plates. One is the North American Plate. The other is the Pacific Plate. These two plates are slipping past each other. Along this fault, earthquakes are very common.

Section Review

Main Ideas: Scientists think the earth's crust is broken into about ten plates. The plates can collide, move apart, and slip past one another.

Questions: Answer in complete sentences.
1. How is the crust of the earth different from the skin on a fresh apple?
2. What are three kinds of movement that occur when plates of the crust meet?
3. How are earthquakes and volcanoes related to plates of crust?

The picture shows a fossil of a sea animal. It lived a long time ago. When the animal was alive, it lived in the ocean. The fossil was found on top of a mountain. How did a sea animal get to the top of the mountain? When you finish this section, you should be able to:

Making Mountains

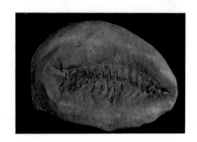

☐ **A.** Identify three changes in the earth's crust that make mountains.

☐ **B.** Describe how the shape of a mountain is the result of how it was formed.

The movement of crustal plates can cause mountains to form in three ways. One way is when two plates collide. When this happens, the crust is bent and squeezed. Squeezing causes the layers of rock to fold. These folded rocks form **folded mountains**. The Appalachian Mountains in the eastern United States are *folded mountains*. The Alps in Europe and the Himalayas in Asia are also folded mountains.

Folded mountains: Mountains formed by folding and squeezing the earth's crust.

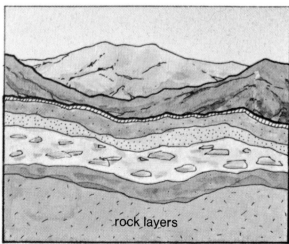

rock layers

**Fault-block
mountains:** Mountains
formed by the
pushing up of rocks
along a fault.

When rocks are squeezed together, they do not always bend. They may crack and tilt. When this happens, mountains are formed. These mountains have large blocks of rock divided by faults. They are called **fault-block mountains**. The Sierra Nevadas, in the western part of the United States, are *fault-block mountains*.

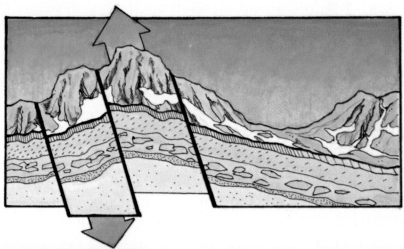

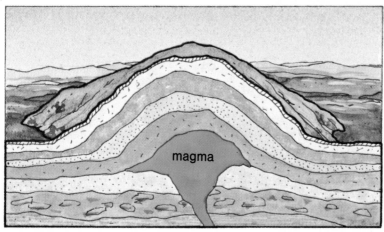

Magma may force its way under layers of rock. It pushes in between the layers, and bends the layers above. This action in the crust causes **dome-shaped mountains**. Stone Mountain, near Atlanta, Georgia, is a *dome-shaped mountain*.

Dome-shaped mountains: Mountains formed when magma pushes up part of the crust, without breaking the surface.

37

ACTIVITY

How are mountains formed?

A. Gather these materials: clay, 2 blocks of wood, a golf ball, and a plastic knife.

B. Make 3 stacks of clay, each with 3 thin layers. Pretend each layer of clay is a rock layer.

C. Hold 1 stack of clay between 2 blocks of wood. Hold 1 block of wood steady and press the clay layers together with your other hand as hard as you can.
1. What happened to the clay?
2. What kind of mountain did you make?

D. Make a cut in a second stack of clay with a knife, as shown in the picture. Hold one half of the stack steady. Push the other half along the cut in the stack.
3. What happened to the clay?
4. What kind of mountain did you make?

E. Hold the third stack of clay in your hand. Gently push a golf ball up under the clay. Remove the golf ball. Cut the stack of clay in half. Look at the layers of clay.
5. What happened to the clay?
6. What kind of mountain did you make?

F. Make a drawing of each stack of clay. Label the name of each type of mountain that you made.

Remember the fossil shown on page 35? The animal lived in the ocean. But the fossil was found on top of a mountain. Can you now explain why?

As you saw in this section, movements of the crust form mountains. The ocean floor was pushed up to become a mountain. The dead animal lay on the bottom of the ocean. When the crust moved, the animal was moved, too.

Section Review

Main Ideas: Movement in the earth's crust can make mountains. Folded mountains form when plates collide. This causes the rocks to fold and bend. Fault-block mountains form when blocks of crust are pushed up along a fault. Magma forces its way in between layers of rock to form a dome-shaped mountain.

Questions: For numbers 1–3, match the cause with its effect. Answer number 4 in a complete sentence.

Cause	Effect
1. Tilting of rocks along faults	a. Dome mountain
2. Magma forces its way under rocks	b. Folded mountain
3. Squeezing of layers of rock	c. Fault-block mountain

4. What could have caused the sea animal fossil on page 35 to end up on top of a mountain?

CHAPTER REVIEW

Science Words: Match the words listed in column A with the definitions in column B.

Column A	Column B
1. Pangaea	a. Large sections of the earth's crust
2. Continental drift	b. Mountains formed when magma pushes up rock
3. Fossils	c. Mountains formed by squeezing of rock layers
4. Plates	d. The idea that the continents are moving
5. Folded mountains	e. Name of earth's super-continent
6. Fault-block mountains	f. Mountains formed when rocks move along a fault
7. Dome-shaped mountains	g. Traces of plants and animals in rocks

Questions: Answer in complete sentences.

1. How do the shapes of the continents help explain that the continents moved?
2. Why are earthquakes and volcanoes found in the same places?
3. What kinds of forces cause mountains to form?
4. What happens at the edges of two plates when they collide?
5. Imagine traveling in a time machine into the future. Would the earth look the same millions of years from now?

CHAPTER 3

THE CRUST WEARS AWAY

The picture above shows how the Appalachian Mountains might have looked when they were young. The picture on page 42 shows the Appalachians today. Have you ever thought of mountains getting old, like people? What happens to a

3-1.

Breaking Down Rocks

mountain as it ages? When you finish this section, you should be able to:

☐ **A.** Describe four things that cause rock to break down into smaller pieces.

☐ **B.** Group the ways in which rocks break down as either *chemical* or *physical weathering*.

What would happen if you hit a large rock with a hammer? The rock would probably break into small pieces. Nature has its own way of breaking rocks and mountains. It is called **weathering** (**weather**-ing). Small pieces of broken rock can be found in most places. *Weathering* helps to change the surface of the earth.

Did you know that water can cause rocks to weather? Water drips into cracks in a rock.

Weathering: The breaking of rock into smaller pieces.

42

When the water freezes, it expands. The rock cracks and splits even more. This kind of weathering is called **physical weathering** (**fizz**-eh-kahl). As the rock breaks into small pieces, only its size and shape change. The **minerals** (**min**-err-als) that the rock is made of do not change.

Have you ever seen a sidewalk cracked by the roots of a tree? Plants also cause *physical weathering*. Plant roots work their way through small cracks in a rock. As the roots grow, they break the rock into smaller pieces. Sandblasting to clean stone buildings is like physical weathering.

Another kind of weathering breaks down rocks by changing the *minerals* in the rock. **Chemical weathering** (**kem**-eh-kahl) changes, adds to, or removes a rock's minerals. Rocks that contain iron can turn red. This is because the iron rusts.

Rusting occurs when iron and water come in contact with each other. What examples of rusting have you seen around your home? A bicycle left out in the rain often will start to rust. The metal bike has iron in it. Rainwater, air, and iron mix to form rust.

Rocks can also be weathered by **carbon dioxide** (**kar**-bon die-**ox**-ide) and water. *Carbon dioxide* is a gas in the air. It mixes with rain and falls on rocks. The mixture of carbon dioxide and water makes a weak acid. The rocks are slowly worn away by this acid. The acid changes the minerals that the rock is made of. This is a kind of *chemical weathering*.

Physical weathering: The changing of a rock's size and shape as it breaks down.

Minerals: Materials of which rocks are made.

Chemical weathering: A change in the minerals of a rock as it breaks down.

Carbon dioxide: A gas in the air that mixes with rain to weather rocks.

43

ACTIVITY

How can rocks be changed by weathering?

A. Gather these materials: marble chips, 2 jars, water, and white vinegar.

B. Fill each jar halfway with rocks.

C. Cover the rocks in one jar with water. Cover the rocks in the other jar with vinegar.

  1. What happens in each jar?

D. Set both jars aside for 24 hours. Look at the rocks in each jar.

  2. Have the rocks in the water been changed?

  3. How have the rocks in vinegar changed?

  4. Is this change an example of chemical weathering? Why?

Section Review

Main Ideas: Weathering breaks rocks into smaller pieces. Physical weathering only changes the rock's size and shape. Chemical weathering changes the minerals in the rock.

Questions: Answer in complete sentences.

1. What could happen if the seed of a tree started to grow in a small crack of a rock?
2. What is the difference between physical weathering and chemical weathering?
3. How do rocks change when water in them freezes?

Rain is not always good. Heavy rain causes rivers to flood. The rising water can cause much damage. What effect do you think a flood like the one shown has on soil, rocks, houses, and trees? When you finish this section, you should be able to:

☐ **A.** Explain *erosion* and how it can be stopped.

☐ **B.** Describe how water moves rocks and soil from one place to another.

What happens to rainwater when it falls to the ground? Some of it soaks into the soil. Some of it is used by plants. But much of the rainwater runs off along the ground. This runoff carries away soil and other loose material. The movement of rocks and soil by rainwater is a kind of **erosion** (ee-ro-zhun).

Erosion: The carrying away of rocks and soil by wind and water.

During heavy rains, a lot of soil can be removed by runoff. It takes a long time to replace the lost soil. Nature needs about 1,000 years to make 2 1/2 centimeters (1 inch) of soil. If the soil is washed away, farmers cannot grow the plants we eat. So people try to prevent soil *erosion*. Erosion can be stopped by planting grass and trees. Plowing across hillsides cuts down erosion. The plowed fields stop the water from flowing straight down the hill.

Soil is first carried away by small streams. The streams then carry the soil to rivers. One river may join other rivers. Let's see how rivers and river systems change the surface of the earth. Rivers flow downhill. They carry along soil and

pieces of rock. These pieces of rock hit and loosen other rocks along the sides of the rivers. Rivers erode the land. After a long time, rivers can cut very deeply into rock. Look at the picture of the Colorado River flowing through the Grand Canyon. How do you think the Grand Canyon was formed?

What happens to the soil that rivers erode? The picture above shows the Russian River flowing into the ocean. The soil and rocks carried by the river are called **sediment** (**sed**-i-ment). When the river reaches the ocean, the *sediment* drops to the bottom. Over many years, the sediment piles up. This forms a piece of land called a delta.

Sediment: Broken-up rock carried in a river.

ACTIVITY

How much rock can moving water carry?

A. Gather these materials: large jar with screw lid, water, and a handful of small pebbles.

B. Put the pebbles in the jar. Fill the jar with water. Screw the lid on tightly.

C. Imagine that the water in the jar is a river. Swirl the "river" in the jar quickly.
   1. What happens to the pebbles?

D. Swirl the "river" slower.
   2. What happens to the pebbles?

E. Swirl the jar until the water is barely moving.
   3. Where are the pebbles now?
   4. How does the speed of the water affect the number of pebbles that are held up?
   5. How does the speed of a flowing river affect the amount of sediment carried?

Section Review

Main Ideas: Moving water can carry away rocks and soil. Runoff erodes the land by carrying soil and broken rock away. The sediment in rivers is carried to the ocean.

Questions: Answer in complete sentences.

1. What happens to rain when it falls to the ground?

2. What happens to the soil and rock that rivers carry away?

3. What is erosion?

The Houston Astrodome is hundreds of feet high. On the mountains of the world are chunks of snow and ice that are even larger than the Astrodome. Imagine how heavy these pieces of ice are! They press down on the mountain slopes below them. They can change the surface of the earth's crust. What are these giant pieces of ice called? When you finish this section, you should be able to:

☐ **A.** Explain how a *glacier* forms.

☐ **B.** Describe four ways that *glaciers* change the crust of the earth.

About 12,000 years ago, North America and Europe were much colder than they are today. Much of the snow that fell did not melt. Year after year, the snow piled up. Over 3 km (2 miles) of snow piled up. It pressed down the snow below it to make ice. How did this huge pile of ice change the surface of the land?

Have you ever made a snowball? What happens when you squeeze the ball? The snow near the center turns to ice. The same thing happened during the Ice Age. The snow near the bottom was pressed together as more snow fell. Slowly, the snow at the bottom turned to ice. The body of ice and snow became heavy. It began to move. A large body of moving ice and snow is called a **glacier** (**glay**-shur). *Glaciers* flow downhill from mountains to lower ground. They erode the land by breaking rocks and carrying them away.

Glacier: A large body of moving snow and ice.

Look at the picture of the mountain. It has very steep sides and a pointed top. This mountain was eroded by glaciers. Its rocks were pulled out by the moving ice. This mountain is called the Matterhorn. It is in the Alps. Glaciers were first studied in the Alps.

Glaciers often push rocks out in front of them. Large rocks, called boulders, are carried on the top of glaciers. Rocks stuck in the bottom and sides of glaciers act like claws. They scrape, scratch, and dig into the crust of the earth as the glacier drags them along.

As the earth got warmer, most of the glaciers melted. We can still see the changes caused by glaciers. Where a glacier pushed rocks in front of it, we now find long, low hills. Large boulders can be found in strange places. The ice carried them from far away. Look at picture 1. As the ice melted, the boulders dropped to the ground. In other places, we find scratches in rock. They were caused by the glacier. Can you see the scratches in picture 2? Look at the map. It shows the glaciers of the last Ice Age in North America. Was the place where you live once covered with ice? The glaciers moved down from Canada. They reached as far south as the present Ohio and Missouri rivers and central Long Island.

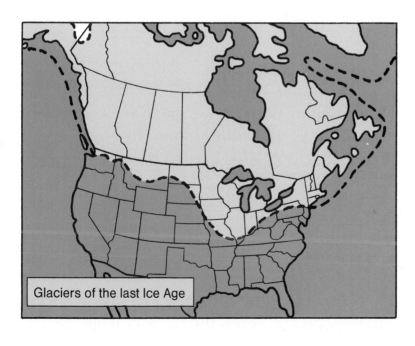

Glaciers of the last Ice Age

ACTIVITY

How do glaciers change the landscape?

A. Gather these materials: 2-liter plastic container with a tight lid, gravel, and water.

B. Put a big handful of gravel in the container. Fill it with water. Close the lid tightly.

C. Put the container in a freezer overnight.

D. When the water is frozen solid, remove the ice from the container.

    1. In what ways is the ice like a glacier?

E. Take the ice outdoors. Lean on the ice as you push it over the ground.

    2. How does pressing down on the ice make it act like a glacier?

    3. How did the ice and gravel mixture change the ground?

    4. Would the same changes have happened without the gravel?

Section Review

Main Ideas: Glaciers are giant masses of snow and ice. As they flow downhill, they erode mountains, scrape rocks, and move loose rocks.

Questions: Answer in complete sentences.

1. How is the formation of a glacier like the making of a snowball?

2. As a glacier moves over the land, what happens to the ground under the glacier?

3. Why do you think the glaciers melted?

Did you know that some of the largest lakes in the United States were made by humans? The lake shown in the picture was once part of a river. Norris Dam was built on the Tennessee River. This trapped the river water. A lake was formed. Building dams is one way people change the earth's surface. When you finish this section, you should be able to:

Humans Cause Change

☐ **A.** Identify ways people change the surface of the earth.

☐ **B.** Describe the effects of changes on the land.

The pictures on the next page show ways humans change the land. Let's look at these more closely to see how we change the land.

Picture 1 shows how humans speed up erosion. This giant shovel removes tons of rocks. The rocks contain iron ore. Steel is made from iron ore. It is used to make such things as cars, airplanes, and rockets. Humans remove rocks and soil from the land just as rivers and glaciers do.

Picture 2 is an example of how humans help to save the land. It is a picture of a national park. Signs in national parks often say, "Take only pictures, leave only footprints." These signs remind people that parks have a special purpose. Parks like these save the homes of many plants and animals. Rivers and mountains are saved. Humans can enjoy the park's beauty.

You learned how nature builds up the surface of the land. Mountains and volcanoes are formed by changes in the earth's crust. Picture 3 shows how humans build on the land. We build cities,

houses, factories, highways, and piles of garbage. All this building changes the land.

Building can cause erosion of land. Removing plants can cause soil to be washed away. What would the land look like if plants were not returned to it? By planting trees and grass, humans prevent erosion.

Humans can work with nature to bring about helpful changes. For example, humans do not cut down all the trees in a forest. Only the older trees are cut. The smaller trees are left behind. They prevent erosion as they grow larger. Often, too many trees are cut down. Humans can replace these trees with young trees. This will help prevent erosion. It will also speed up the growth of the forest.

Humans have learned that planning is important. Before people change the land, they need to think about the effects the change will have.

ACTIVITY

How do humans change the land?

A. Gather these materials: old magazines, poster paper, scissors, and glue.

B. Look at the magazines. Find pictures of humans changing the land. Find at least 10 pictures.

C. On the poster paper, put the pictures into three groups:

Group 1: How humans erode the land.

Group 2: How humans build on the land.

Group 3: How humans help to save the land.

D. Paste the pictures on the poster paper.

E. Draw a star next to the pictures that show changes that will last for a long time.

Section Review

Main Ideas: Humans change the land in many ways. Removing rocks and minerals from the earth's crust wears down the land. Building roads, factories, and houses changes the appearance of the land. Humans can save the land.

Questions: Answer in complete sentences.

1. Look at picture 1 on page 54. What type of change have humans caused?

2. What are some ways humans change the land? Find pictures in this book for each change you list.

3. What would be the effect of building apartments in wooded areas?

CHAPTER REVIEW

Science Words: Match the words in column A with the definitions in column B.

Column A	Column B
1. Physical weathering	a. Materials rocks are made of
2. Weathering	b. A large body of moving snow and ice
3. Erosion	c. Carrying away rocks and soil
4. Carbon dioxide	d. Changing the size and shapes of rocks
5. Chemical weathering	e. Changing the minerals in a rock
6. Glacier	f. Material carried by rivers
7. Minerals	g. Breaking rocks into smaller pieces
8. Sediment	h. Mixes with rain to weather rocks

Questions: Answer in complete sentences.

1. How can water and plants weather rocks?
2. How is physical weathering different from chemical weathering?
3. How does rain cause erosion of the earth's crust?
4. What clues on the land show that a glacier passed over it?
5. Listed below are changes humans have made to the land. Classify each as (a) building up the land, (b) wearing down the land, or (c) not changing the land.

   1. Building a tunnel
   2. Making a state park
   3. Building a skyscraper
   4. Farming the land

How do plants affect erosion?

A. Gather these materials: 2 large aluminum roasting pans, soil, sod, sprinkling can, hammer, nail, 2 jars, and water.

B. Use the hammer and nail to make a 1-cm-wide hole in the bottom of each pan. The hole should be near the pan's end.

C. Cut a piece of sod to fit in one pan. Fill the other pan with soil. Make the soil the same depth as the sod.

    1. What is different about the 2 "hills"?

    2. What do you think will happen to the surface of these "hills" when you pour water on them?

D. Fill the sprinkling can with water. Sprinkle water over the top of each "hill." Hold a jar under the hole at the bottom of each "hill."

    3. How did the water affect the surface of each "hill"?

    4. How does grass on a hillside affect erosion? Why?

Seismologist ►

You have learned that earthquakes cause a lot of damage. It is important to know about earthquakes before, during, and after they happen. Seismologists are people who study earthquakes. They use a machine called a seismograph to measure earthquakes. People who are seismologists have studied geology and math in college.

◄ Surveyor

You may have seen surveyors working near your home. A surveyor uses machines to measure the high and low places of the land. Surveyors can also measure the size and shape of pieces of land. They usually work in teams. A surveyor needs some background in mapmaking and math. Some surveyors are trained on the job.

59

LIGHT

UNIT 2

CHAPTER 4

LIGHT BEAMS AND SHADOWS

4-1.

Light Sources

Pretend that you are sitting around an open fire on a camping trip. The moving flames cast shadows that dance all around you. The shadows seem spooky. There is another kind of light shown in the picture. What is the source of this light? When you finish this section, you should be able to:

☐ **A.** Identify objects that are light sources.

☐ **B.** Explain how we see objects that do not give off their own light.

What do fireflies, candles, the sun, light bulbs, and fireworks have in common? The answer is that they all give off light. Each of these objects, even the firefly, has the ability to make light. We call objects such as these **light sources**. Visible light is a kind of energy that we are able to see with our eyes. *Light sources* help us see things that do not give off their own light. Without them, we would live in a completely dark world.

Light sources: Objects that produce light.

Light sources include the sun, stars, light bulbs, headlights on cars, and matches. Light sources make their light in different ways. When a match gets hot enough, it starts to burn. The burning wood gives off light. Stars, like the sun, are balls of hot gases. Fireflies have chemicals in their bodies that make light without heat.

Light energy includes the light you see and also light you can't see. Have you heard of X rays, radio waves, and microwaves? These are forms of light that our eyes are not able to sense. When you have an X ray taken of a broken bone, the doctor shines a form of light at you. X rays are able to pass through your body and hit a piece of film. A picture of your bones appears on the film.

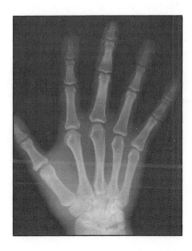

It's true that you can do many things. But no matter how hard you try, you cannot give off visible light. The book you are reading does not give off light. Then, how can other people see you? How can you see the pages of this book? Light sources, like the sun or a light bulb, make objects visible. Let's see how this happens.

When light hits an object like this book, some of it bounces off. The light is bounced off the book to your eyes. This is called **reflection**. In this way, we are able to see objects that do not give off their own light. Imagine there are no windows in the room in the picture. Would the girl be able to see objects in the room if the light went out?

Reflection: The bouncing back of light.

We see most objects by *reflected* light. The moon is a good example. On some nights, the moon seems to be shining very brightly. Is it giving off its own light? The moon reflects light from the sun.

ACTIVITY

Can we see a light beam?

A. Gather these materials: a flashlight and 2 used chalkboard erasers.
B. Work with a partner. Darken the room. Turn the flashlight on and aim it across the room.
   1. Can you see the light beam?
C. Have your partner walk along the path of the light. Have your partner gently tap 2 chalkboard erasers together above the light beam.

   2. What do you see now?
   3. How did the chalk dust help you see the beam of light?
   4. Can light be seen when there is nothing to reflect it?

Section Review

Main Ideas: Light is a form of energy. Light sources are objects that make their own light.

Questions: Answer in complete sentences.

1. Make a chart with the words "Light Source" and "Reflector of Light" at the top. List five objects that belong in each group.
2. How is a light source different from an object that reflects light?
3. The earth and Mars are planets. A planet does not give off its own light. Why can we see planets?

4-2.

Straight Lines

Can you imagine riding on a beam of light? It sounds like a crazy idea, doesn't it? Albert Einstein (**ine**-stine), a famous scientist, wondered what this would be like. He often used his imagination to try to find answers to his questions. Light travels very fast. If you could travel around the earth at the speed of light, you would finish seven orbits in only 1 second! If you went on a round trip to the sun on a beam of light, it would take only 17 minutes. In the space shuttle, it would take you 155 days! When you finish this section, you should be able to:

☐ **A.** Compare the speed of light with the speeds of other objects.

☐ **B.** Describe how shadows are formed.

Light travels very fast. Blink your eyes. You probably blink your eyes once in a second. During the time that it takes you to blink once, a beam of light can travel 300,000 kilometers (186,000 miles)! To understand how fast light travels, consider the following example. A light beam can reach the moon and be reflected back to earth in less than three seconds. A trip to the moon and back in a rocket would take several days.

Light travels in straight lines. How do we know this? First, we cannot see "around the corner" of an object. Second, when light hits an object, the object casts a sharp **shadow**. A *shadow* is an area of darkness formed when light is blocked by an object. Shadows help us understand that light travels in straight lines. Let's look at some examples.

Shadow: The dark area caused when an object blocks light.

Look at the picture of the candle and the shadows. What do you see that shows that light travels in straight lines? How would it look if the light from the candle curved around the pencils? The shadows would disappear. Since there are shadows, the light must be traveling in straight lines.

Look at the shadows formed when a bright light hits these objects. The light that hits the object is blocked. Since the shadow is the same shape as the object, the light must be traveling in straight lines. If it were not, the shadows would not be the same shape as each object. The shadows would be blurred around the edges.

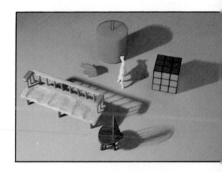

Does light travel in straight lines?

A. Gather these materials: 4 index cards, hole punch, clay, and light source.

B. Punch a hole as close as you can to the center of each card.

C. Use a piece of clay to stand the card up on a desk. Turn on the light and put the lamp at the other end of the desk. Look through the hole at the light.

D. Line up the next card so that you can see the light through both holes. Continue until you can see the light through the holes in all 4 cards.

   1. Are the holes all in a straight line?

E. Move 1 card to the left.

   2. Can you still see the light through the holes?

   3. What does this tell you about how light moves?

Section Review

Main Ideas: Light travels in straight lines at a very high speed. When objects block the path of light, shadows are formed.

Questions: Answer in complete sentences.

1. Imagine you could travel at the speed of light. How long would it take you to travel (a) to the moon; (b) to the sun; (c) from school to home?
2. How is a shadow formed?
3. What would be different if light did not travel in a straight line?

People in Science

Harold E. Edgerton

Harold Edgerton was born on April 6, 1903, in Fremont, Nebraska. Dr. Edgerton developed a method of high-speed photography in 1931. At the Massachusetts Institute of Technology, he invented an electric lamp called the *electronic flash*. This lamp made a very bright light. It could be flashed on and off very rapidly. Using this flash, Dr. Edgerton was able to take "stop-action" pictures. Such pictures have been very helpful to scientists. One series of pictures shows what happens when a drop of milk falls into a saucer of milk. Another shows the moment at which a bat hits a baseball. Dr. Edgerton has also done important work in aerial and underwater photography.

4-3.

Spreading Light

A lighthouse can be a beautiful sight. Imagine you are the captain of a ship. You are lost at sea, but you keep heading east. Ahead you see a faint light. A little while later, the light is brighter. It's a lighthouse. Why does the light on the lighthouse get brighter? When you finish this section, you should be able to:

☐ **A.** Describe the way light travels away from its source.

☐ **B.** Describe what happens to light as it gets farther away from its source.

☐ **C.** Describe how light acts as either a particle or a wave.

You learned in the last section that light travels in straight lines away from its source. As light travels away, it spreads out in all directions. In a way, the light gets dimmer. When the source of light is very far away, only a few light beams reach our eyes. The farther away the light source is, the dimmer the light seems.

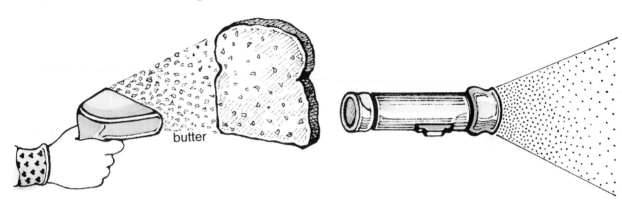

butter

To understand this idea, look at the picture of the "butter gun." The butter spreads out as it leaves the gun. If you hold the gun far away, you can butter a lot of toast at once. But the amount of butter on each piece is very thin. Think of the butter gun as a light source like a flashlight.

Scientists have two ways of explaining how light acts. One idea is that light acts as if it were made of tiny pieces of matter, or **particles** (par-tih-kulz). Have you ever used a spray bottle to spray water on plants? Little *particles* of water come out of the nozzle. When the water drops leave the nozzle of the bottle, they are close together. The water drops spread apart as they travel away from the nozzle. Light acts in the same way. Why do you think the light is dimmer farther from a flashlight?

Particle: A very small piece of matter.

71

ACTIVITY

How does light spread out from a source?

A. Gather these materials: centimeter graph paper, flashlight, and meter stick.

B. Place the graph paper on a flat surface. Each square measures 1 centimeter (0.4 in.).

C. Turn on your flashlight. Hold it close to the graph paper. The bright circle of light should just fill one 1-cm square. (If your light is too big, try to fill the smallest number of squares you can.)

D. Measure the distance from the flashlight to the graph paper.

E. Keep moving the flashlight away from the graph paper until the light is too dim to see. Record all your distance measurements.

1. How does this activity show that light spreads out as it gets farther from the source?

2. As the flashlight gets farther from the graph paper, what happens to the brightness of the light in each square?

3. How far must the flashlight be from the graph paper before the light is too dim to see?

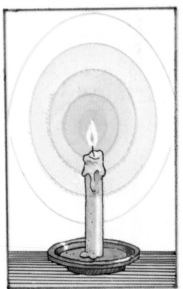

Have you ever dropped a stone in water? When the stone hits the water, **waves** travel away in all directions. Scientists think light also acts like *waves*. Light travels away from the source as a wave. The light wave gets weaker as it travels. Since the wave gets weaker, the light appears dimmer.

Wave: A movement like a swell of water.

Section Review

Main Ideas: As light moves away from its source, it appears to get dimmer. This happens because light spreads out as it travels. Scientists have observed that light acts like both a particle and a wave.

Questions: Answer in complete sentences.
1. Why does light from a street lamp get dimmer as you move away?
2. How does light behave like particles?
3. How does light behave like waves?

4-4.

When Light Strikes

These mountain climbers are crossing a snow-field. The sunlight is bright enough to hurt their eyes. The climbers are wearing sunglasses. How do the glasses protect their eyes? When you finish this section, you should be able to:

☐ **A.** Explain three things that can happen to light when it hits an object.

☐ **B.** Group objects by what happens when light hits them.

What happens to light when it hits an object? Light passes right through some objects. But it cannot pass through other objects. Let's find out how light behaves when it strikes different kinds of objects.

 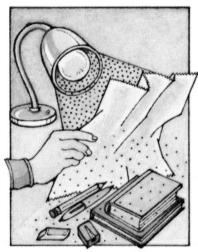 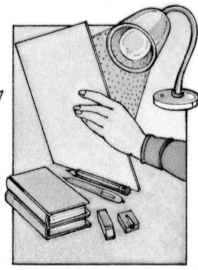

Some materials like glass, water, and air let light pass right through. If a flashlight is held on one side of an aquarium of water, the light can be seen on the other side. Objects through which light passes easily are called **transparent** (tranz-**pair**-ent). You can see through *transparent* objects clearly.

Sometimes when light passes through an object, things look blurred. Have you ever tried to look through a piece of waxed paper? When you do, things on the other side are not very clear. Some of the light goes through, but not enough to see clearly. The rest of the light is either bounced away or soaked up. Materials like waxed paper are called **translucent** (tranz-**loo**-sent). Think of other *translucent* materials?

What would you see if you tried to look through a piece of cardboard? You would see nothing but the cardboard. Light does not pass through some materials. Materials that block light are called **opaque** (oh-**payk**). Do you think a piece of aluminum foil is *opaque*?

Transparent material: A material through which light can pass.

Translucent material: A material that blurs light as it passes through.

Opaque material: A material that blocks light.

ACTIVITY

What happens when light hits objects?

A. Gather these materials: a flashlight and a variety of materials such as glass, plastic, paper, fabrics, tape, wood, and metal.

B. Copy the chart shown here.

Materials	Opaque	Transparent	Translucent
1. Wood			
2.			
3.			

C. Shine the light at each material. Record your results in the chart.

Section Review

Main Ideas: Objects affect light in three ways.

	What Light Does	Object
Opaque	It is blocked	Paper
Transparent	Passes through	Glass
Translucent	Scatters	Waxed paper

Questions: Answer in complete sentences.

1. Is air transparent, translucent, or opaque?
2. Are sunglasses transparent, translucent, or opaque?
3. How do sunglasses protect the mountain climbers' eyes in the picture on page 74?

CHAPTER REVIEW

Science Words: Write the meaning of each of the following words. Then give at least two examples of each word: (1) light source; (2) reflection; (3) shadows; (4) particle; (5) wave.

Questions: Answer in complete sentences.

1. Why can we see the moon? Explain your answer with a diagram.

2. Draw and label a diagram to show how a shadow is made. Include in your diagram a light source, an object, and the shadow.

3. Do transparent, translucent, or opaque objects make the best shadows? Why?

4. Look at the picture of two boats and a lighthouse. Would the light seem brighter to the captain of boat A or boat B? Explain your answer.

5. What will happen to light when it hits the following objects: (a) clear plastic; (b) a piece of wood; and (c) waxed paper.

6. What is the difference between a transparent and a translucent material? Give an example of each.

CHAPTER 5

BOUNCING LIGHT

5-1.

Reflecting Light

In ancient Egypt, people used pieces of polished metal as mirrors. Why do some materials reflect light so perfectly? What kinds of materials make good mirrors? When you finish this section, you should be able to:

□ **A.** Describe how smooth and rough surfaces reflect light in different ways.

□ **B.** Predict which materials will make good reflectors.

Some materials reflect light better than others. Good reflectors are usually shiny and bright when light is shined on them. If you look at a good reflector, you may see yourself clearly. If you look at a poor reflector, your reflection may be blurred or not there at all.

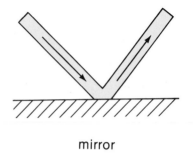

mirror

Imagine that you and your friends have a basketball. Where would you rather bounce the ball, on a gym floor or on a dirt road? It is easier to bounce a ball on the gym floor because the floor is smooth.

Think of a beam of light as a stream of tiny particles. In the diagram, the beam of light hits the flat surface of a mirror. The beam bounces off the mirror like a basketball would bounce off a smooth gym floor.

What happens when light hits an uneven surface? The particles of light are scattered. When this happens, the light is not reflected evenly.

Can you predict a good reflector?

A. Gather these materials: flashlight, mirror, 2 squares of aluminum foil, and piece of white paper.

B. Crumple 1 piece of aluminum foil until the surface is uneven. Flatten the foil out.

C. Use the mirror and each piece of aluminum foil to read the secret message.

   1. With which materials were you able to read the message?

   2. Which piece of foil makes a better reflector? Why?

D. Hold the mirror in one hand and the flashlight in the other. Aim the flashlight at the mirror so you bounce a reflection off the ceiling. Repeat this step with both pieces of foil and white paper.

   3. With which materials were you able to bounce light off the ceiling?

   4. What makes an object a good reflector?

FLAT SURFACES
MAKE GOOD
MIRRORS

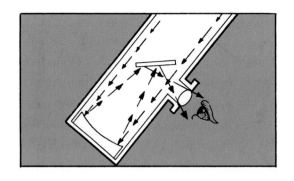 

A good mirror is a very smooth reflector. Reflectors are used in cameras, flashlights, telescopes, and microscopes.

The diagram on the left shows how telescope mirrors reflect light from stars and planets. On the right, the diagram shows how light is reflected in a microscope.

Section Review

Main Ideas: When light hits an object, it can be reflected in two ways. A beam of light can bounce off the surface in a straight line. An uneven surface scatters the light particles in many directions. Flat, smooth surfaces are good reflectors. Uneven surfaces are poor reflectors.

Questions: Answer in complete sentences.
1. How is a Ping-Pong ball hitting a table like light hitting a mirror?
2. What is reflection? Give an example of a material that would give a good reflection.
3. If you wanted to make a reflector for your bike, what kind of material would you use?
4. Why do you see a good reflection in a clean mirror?

5-2.

Flat Reflectors

How does the captain of a submarine see what is happening above the water? A submarine uses a long tube that contains mirrors. The light from above the water is reflected down the tube. When you finish this section, you should be able to:

☐ **A.** Explain how a mirror changes the direction of a light beam.

☐ **B.** Predict the direction light will reflect when it hits a mirror.

☐ **C.** Explain how a *periscope* works.

If you were to throw a ball at the ground, it would bounce off the ground. The direction that it bounces depends on how it hits the ground. The direction the ball bounces is called the **angle**. The *angle* will depend on how the ball hits the ground. Light acts this way, too. Light will bounce off objects at different angles.

Angle: The direction at which light bounces off something.

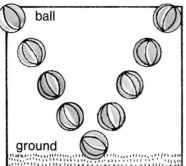

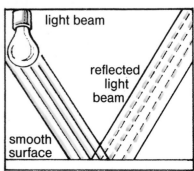

ACTIVITY

How does a mirror affect light?

A. Gather these materials: clay, light box, mirror, and sheet of white paper.

B. Turn on the light box. Place the mirror in the path of the light beam. Move the mirror so that the light hits it at different angles.

    1. What happens to the light beam when it hits the mirror?

    2. How can you use the mirror to change the light's direction?

    3. How can you get the light beam to reflect back onto itself?

C. Bounce a beam of light off the mirror at an angle. Draw a diagram that shows the path of the light beam hitting the mirror. Show the path of the reflected beam.

    4. How does the angle of the light beam compare with the angle of the reflected beam?

    5. How is light affected by a mirror?

Periscope: An instrument that uses two mirrors to see around corners.

Have you ever tried to signal someone by using a mirror? Flat reflectors are used for this. The light from the sun reflects off the flat mirror. The mirror is tilted to an angle. When the angle is just right, the light reflects to the eye of the other person.

Have you ever looked through a **periscope** (**pehr**-ih-skope)? A *periscope* uses two mirrors to "see" around corners. The mirrors are placed so that one mirror reflects light beams to a second mirror. The second mirror then reflects the light beams to your eyes.

A mirror changes the direction of light. When you change the position of the mirror, the direction of the reflected light beam also changes. A light beam can be reflected back onto itself. This happens when the mirror is directly in front of a light beam. A light beam always reflects off the mirror at the same angle as it hits the mirror. But it reflects in the opposite direction. If you want to signal a person, you have to hold the mirror so the light reflects toward the person.

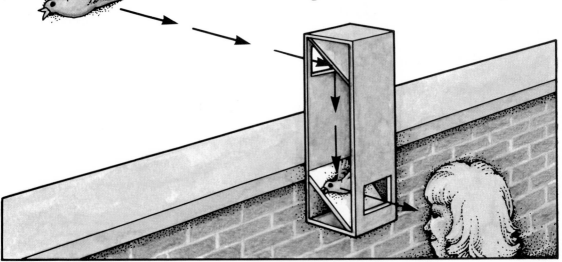

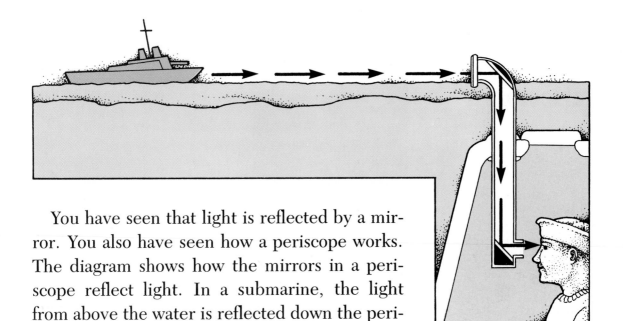

You have seen that light is reflected by a mirror. You also have seen how a periscope works. The diagram shows how the mirrors in a periscope reflect light. In a submarine, the light from above the water is reflected down the periscope. The captain sees the reflection from a second mirror.

Section Review

Main Ideas: Light beams can be reflected by mirrors. Light bounces off a mirror at the same angle that it hits the mirror. A periscope uses two mirrors to see around corners.

Questions: Answer in complete sentences.

1. How is light hitting a mirror like a ball being thrown at a flat wall?
2. How does a periscope help you see around corners?
3. Why do submarines need periscopes?
4. Copy or trace the diagrams. Each line is a light beam hitting a mirror. How will each beam bounce off the mirrors? Draw lines to show your answers.

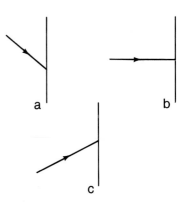

5-3.

Curved Reflectors

Have you ever walked in front of a curved mirror in a museum or fun house? Some mirrors make you look tall and skinny. Other mirrors make you look short and fat. Why do you think this happens? When you finish this section, you should be able to:

☐ **A.** Describe the path of light when it hits a curved reflector.

☐ **B.** Explain how curved reflectors can bring light together.

☐ **C.** Identify ways curved mirrors are used.

Look at the photograph below of the three beams of colored light. The beams are bouncing off a curved mirror. Notice that the three beams cross each other at the same place. The blue beam is reflected back on itself. But the red and green beams hit the mirror at an angle. They are reflected off the mirror at an angle and cross each other. Mirrors with inward curves bring light beams together. These mirrors **focus** (foh-kus) light to a spot. Curved mirrors bend beams of light at many angles.

The diagram shows how the light beams are reflected. You can see very clearly how the beams are *focused* on one spot.

A curved mirror is like many flat mirrors that are turned at angles to make a curve. The diagram on page 89 shows a curved mirror on the left. On the right are several flat mirrors turned slightly to make a curve. In both cases, light is focused to a spot.

Focus: To bring light beams together at a spot.

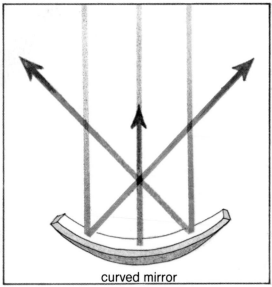

curved mirror

ACTIVITY

How can mirrors be used to focus light?

A. Gather these materials: clay, file card, light box, 4 mirrors, and white paper.

B. Place a sheet of white paper about 40 cm (16 in.) from your light box, as shown.

C. Make a screen by bending the file card. Place it at the side of the white paper.

D. Darken the room. Using the clay, stand a mirror in the path of the beam. Turn the mirror so the light is reflected onto the file card. Place another mirror next to the first one. Turn it so it also reflects light onto the card.

   1. What happened to the brightness of the light when you added the second mirror?

   2. What will happen to the brightness if you add a third mirror?

E. Add a third and a fourth mirror.

   3. How can flat mirrors be used to focus light?

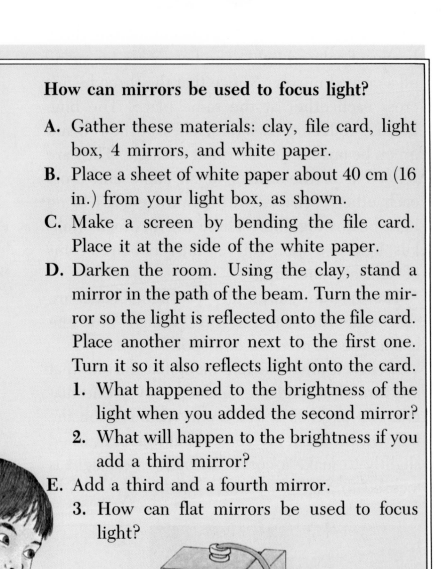

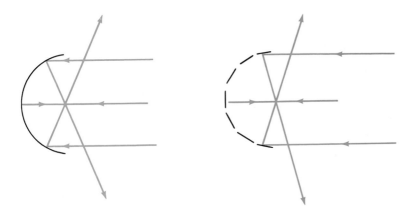

The inside of a spoon is like a curved mirror. If you look at the inside of a spoon, everything reflected in the spoon will look larger. Use a spoon to look at the print on this page. The reflected letters should look bigger. Mirrors like this are called **concave** mirrors.

The outside of a spoon is also a curved mirror. But the mirror curves out. When you look at things with the outside of the spoon, they will look smaller. Mirrors like this are called **convex** mirrors. You have probably seen *convex* mirrors hanging from the ceiling of stores. Why do you think convex mirrors are used for this purpose?

Concave mirror: A mirror that bends in.

Convex mirror: A mirror that bends out.

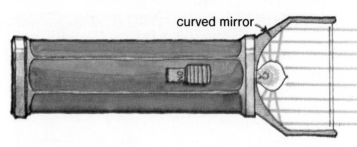
curved mirror

Curved mirrors have many uses. They are used in spotlights, telescopes, flashlights, and the headlights of cars. Curved mirrors focus light. When light is focused, it is brighter. In a flashlight, a bulb is placed at the spot where light is focused by a curved mirror. When you turn the flashlight on, the light from the bulb reflects off the curved mirror. It travels away from the flashlight as a bright beam of light.

Section Review

Main Ideas: When light beams are focused, they are brought together in one spot. Mirrors turned at slight angles can be used to focus light. Curved mirrors are used in flashlights, spotlights, and headlights.

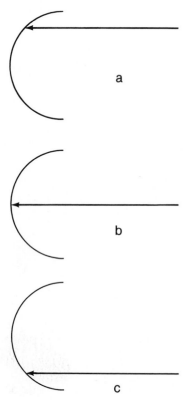

a

b

c

Questions: Answer in complete sentences.

1. What will be the path of the reflected light beams in the diagrams? Draw a diagram to show your answer.
2. Why does your body look bent in a curved fun house mirror?
3. When light hits a curved reflector, what happens to the light?
4. How would a flashlight be different without a curved mirror inside?

CHAPTER REVIEW

Science Words: Match the words in column A with their meanings in column B.

Column A
1. Concave mirrors
2. Convex mirrors
3. Angle
4. Periscope
5. Focus

Column B
a. Used to see around corners
b. Direction of light reflecting off a mirror
c. Mirrors that curve in
d. Mirrors that curve out
e. To bring together at one spot

Questions: Answer in complete sentences.

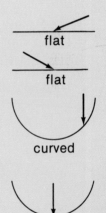

1. Which of the following materials would be good reflectors: (a) polished spoon; (b) mirror; (c) rock; (d) wool sweater; (e) rug; or (f) knife blade?
2. Trace or draw these mirrors. Predict the angles of reflection for the beams of light hitting these mirrors. Draw a line on your diagram to show your answer.
3. What do you think would be the path of light in the bottom diagram?
4. How does a curved mirror focus light? Show your answer with a diagram.
5. What is the purpose of a curved mirror in a flashlight and car headlight?
6. Which would be a better mirror on your bike: convex or concave? Explain.
7. How is a periscope used to see around corners?

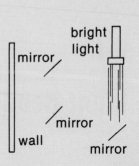

CHAPTER 6

BENDING LIGHT AND COLORS

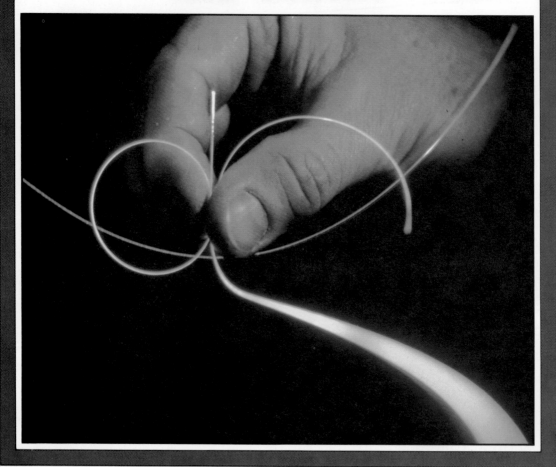

6-1.

Light Bends

Imagine riding in a car on a very hot day. You look down the road and see large puddles of water. As you get closer, the puddles disappear. What are these "puddles of water" called? When you finish this section, you should be able to:

A. Explain what happens to a beam of light when it passes from one kind of material into another.

B. Describe the path of light when it moves from one material into another.

Light not only bounces off materials, it goes through some of them. When light goes from air into a material like water, it bends.

Look at the picture of a pencil in a glass of water. The pencil looks broken. Is this a trick? Light bends when it goes from air to water. Because of this bending, the pencil looks broken.

Because light bends, there are times when light plays tricks on us. Have you ever sat on the edge of a pond and tried to reach for a fish? You probably missed. You missed because the fish was not really where you grabbed. The fish was actually lower than you thought. The light reflected from the fish bends as it moves from water to air.

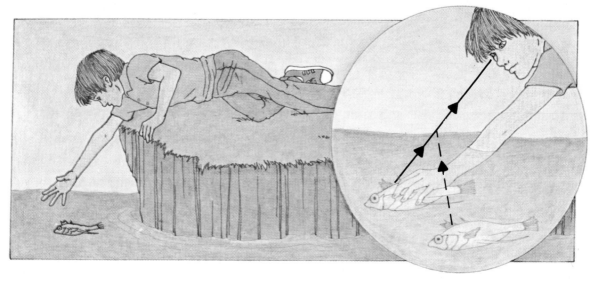

ACTIVITY

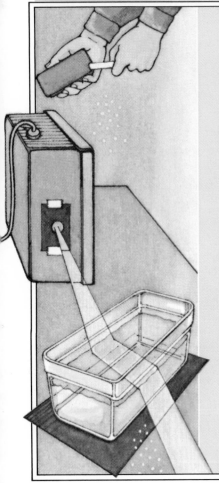

How can you bend a light beam?

A. Gather these materials: clear plastic shoebox, light box, milk, sheet of black paper, water, and used chalkboard eraser.

B. Place the plastic shoebox on the sheet of black paper. Fill the shoebox with water. Mix 4 or 5 drops of milk into the water.

C. Shine the light straight at one long side of the shoebox.

D. Scrape some chalk dust off the eraser. Sprinkle the dust into the light beam. Look down on the light beam from above.

E. Move the light box to a new position.
1. What happens to the path of the light?
2. How can you make the light bend more?
3. How does changing the angle of the light beam affect the path of the light beam in the water?

Mirage: Something we see that really isn't there.

The bending of light explains what seem to be puddles of water on roads. These puddles of water, which are not really there, are called **mirages** (muh-**razh**-ez). On a hot day, the sun heats up the road. This makes a layer of hot air just above the road. The light passing through cooler air above hits the hot air and is bent. The difference between cool and warm air is enough to bend the light as it passes through. When the light bends, you see a reflection of the blue sky.

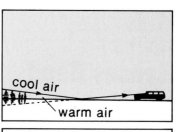

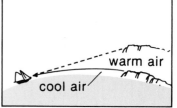

This reflection looks like a puddle of water. As the diagram shows, there are different kinds of *mirages*. Sometimes distant objects like mountains may appear to float in the air.

Section Review

Main Ideas: When light moves from one material into another, its beams are bent.

Questions: Answer in complete sentences.

1. What happens to a beam of light as it passes from one material into another?
2. Look at the picture on page 93. Why does the pencil look broken?
3. Explain one example of how the bending of light can trick your eyes into seeing something that is not really there.

6-2.

Lenses Bend Light

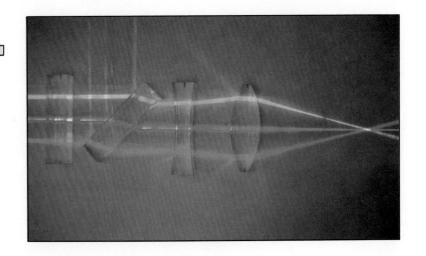

How are air and glass different? Air is a thin material and a gas. Glass is thick and solid. When light passes from air into glass, it is bent. Look closely at the picture. The light is bent so that the beams of light cross each other. Do you know why this happens? When you finish this section, you should be able to:

☐ **A.** Explain what happens to beams of light as they pass through a glass *lens*.

☐ **B.** Identify common objects that contain *lenses*.

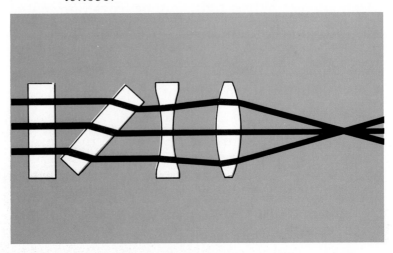

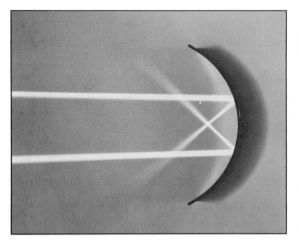
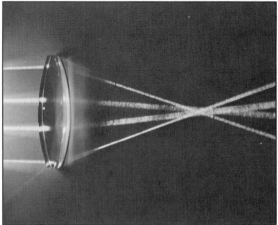

The picture on page 96 shows that light bends when it passes through a **lens** (lenz). A *lens* is any piece of curved glass or plastic. The lens may be curved on one side and flat on the other side. Or it may be curved on both sides. The light is bent as it moves from the air into the lens. The light is bent again as it leaves the lens and enters the air. The point where light beams meet is called the focus or **focal point** (**foe**-kul) of the lens.

Look at the two pictures above. One shows how a curved mirror bends light. The other shows how a lens bends light. Both focus the light to a *focal point*.

Lens: A piece of curved glass that bends light.

Focal point: The point at which a lens brings light beams together.

ACTIVITY

How does a lens work?

A. Gather these materials: 2 jars (1 olive jar and 1 mayonnaise jar), water, flashlight, piece of aluminum foil, sheet of white paper, and ruler.

B. Tape a sheet of white paper to your desk. Draw a straight line across the paper.

C. Fill the narrow jar with water. Place the jar on the line, as shown in the picture.

D. Put the aluminum foil over the end of the flashlight. Cut a narrow slit in the foil. Darken the room and turn on your flashlight. Aim the light at the jar as shown.

   1. What happened to the light as it passed through the jar?

E. Place the wide jar of water on the line as shown. Aim the light at the jar.

   2. What happened to the light as it passed through the jar?

   3. How is light bent by a narrow jar compared with a wide jar?

   4. How are the jars like lenses?

Lenses have many different uses. We can use lenses to get a closer look at small things, like rocks and bugs. Lenses in eyeglasses help people to see better. The lens helps the eye to focus light. This makes the object look sharper.

Lenses are also used in scientific instruments. An instrument used to make very small things look larger is called a **microscope** (**my**-kruh-skope). You can see creatures with a *microscope* that are impossible to see with just your eyes. Some microscopes can make an object look 1,000 times bigger than it really is! Look at the two photos of a housefly. The left picture shows how it looks using only your eyes. The right picture was taken through a microscope. What can you see in the right picture that is impossible to see in the left picture?

Scientists use an instrument called a **telescope** (**tell**-uh-skope) to see things that are far away. *Telescopes* use lenses to magnify objects like

Microscope: An instrument that makes small things look large.

Telescope: An instrument that makes faraway things look closer.

stars and planets. Look at the two photographs. Which one was taken through a telescope? What features of the moon can be seen only with the help of a telescope?

Section Review

Main Ideas: A lens is a piece of curved glass or plastic that lets light pass through. Lenses can bend light beams so that the beams come together at one place. The place where light beams come together is called the focus of the lens. Lenses can make small things look bigger or faraway things seem closer.

Questions: Answer in complete sentences.
1. What happens to light when it passes through a lens?
2. What are three objects that have lenses?
3. What is the same about the way light is bent by a lens and a curved mirror?

Have you ever seen a rainbow? Why do you think rainbows only appear after a rain shower? You can hold a piece of glass in the sun to make your own small rainbow. When you finish this section, you should be able to:

☐ **A.** Explain why a *prism* can split light into colors.

☐ **B.** Identify the colors of the light *spectrum*.

☐ **C.** Describe how a rainbow is formed.

The piece of glass shown below is called a **prism** (**priz**-um). A *prism* breaks white light into a group of colors called the **spectrum** (**spek**-trum). The colors of the *spectrum* are red, orange, yellow, green, blue, indigo, and violet. An easy way to remember these colors is to imagine a boy's name, ROY G. BIV. Each letter stands for one of the colors in the spectrum.

Prism: A piece of glass that bends light and separates it into color.

Spectrum: The group of seven colors that make up white light.

How does a prism split white light into the colors of the spectrum? You have already learned that light can be reflected and bent. Remember what happens when light is bent. When a beam of light passes from air into glass, it slows down. When it passes into the glass, it bends.

The diagram below shows white light hitting a glass prism. When the light hits the glass, it is slowed down and bent. Notice that the color vio-

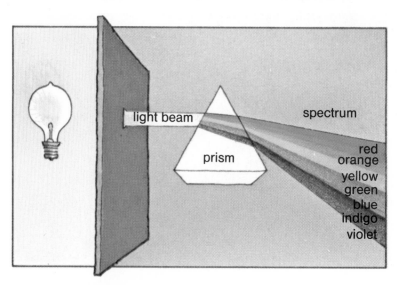

light beam

spectrum

prism

red
orange
yellow
green
blue
indigo
violet

let is bent more than the color red. When the colors come out the other side of the prism, they are bent again. Each color is bent by a different amount. This causes the colors to be separated.

After it rains, there are millions of tiny water drops in the air. When the sun comes out, light strikes the drops of water. Each tiny droplet acts like a prism.

Sunlight enters the drop and is split into the colors of the spectrum. The colors are bounced off the other side of the raindrop. They are reflected to the front of the drop. When they leave the raindrop, the colors are bent again. This bending and reflecting happens in millions of drops at the same time.

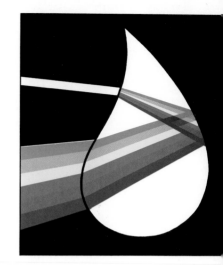

ACTIVITY

What is white light made of?
A. Gather these materials: flashlight, mirror, tray of water, and crayons.
B. Place the mirror in the tray of water.
C. Shine the flashlight at the mirror.
D. Move the flashlight around until you can see a reflection of colors on the ceiling.
   1. What colors did you see?
E. Draw a picture of the spectrum you observed.
   2. What happened to the white light from the flashlight?
   3. What colors make up white light?

103

Isaac Newton was a scientist who split light by using a prism. He held a glass prism in the path of a narrow beam of sunlight. He saw that light is broken into seven colors by a prism.

Section Review

Main Ideas: Prisms can separate white light into a group of colors called the spectrum. The spectrum is made of red, orange, yellow, green, blue, indigo, and violet. Rainbows are a special kind of spectrum. You can sometimes see a rainbow after a rain shower. Each raindrop bends and reflects sunlight like a prism.

Questions: Answer in complete sentences.

1. Explain what a prism does to white light. Draw a diagram to show your answer.
2. What are the colors that make up white light?
3. How are raindrops like a prism?
4. What happens when a prism is placed in front of a spectrum of colors?

What colors do you see in the figure below? Cup your hands around the picture. Bring your eye close to the page and look at the picture now. Can you see colors in dim light? Can you see colors with no light? When you finish this section, you should be able to:

☐ **A.** Explain how we see different colors.

☐ **B.** Describe what happens when white light hits colored glass.

☐ **C.** Identify the primary colors of light.

When white light strikes an object, all the colors of white light also strike that object. If all the colors are reflected, the object will look white. For example, look at a piece of white paper. The

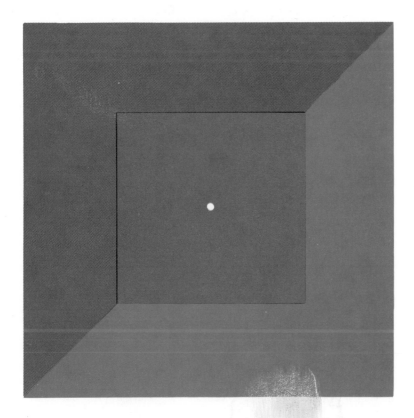

Absorb: To take in and hold light.

paper is white because all the colors are reflected to your eye.

Not all objects are white. Colored objects soak up some of the light that strikes them. They **absorb** (ab-**sorb**) some light. Some of the light is reflected. The apple is red because it reflects the red part of white light. The apple *absorbs* all the other colors. Grass is green because it reflects the green part of the white light. All the other colors are absorbed by the grass. What color is the shirt or dress you are wearing today? Why is it that color?

Some objects allow light to pass through them. Colored glass, cellophane, and colored water are examples. Why do these objects have color?

Look at the picture of the stained glass. When light passes through colored glass, some of the colors in the white light are absorbed. Red glass is red because all the colors except red are absorbed. Only the red part of light passes through. A transparent material lets its own color pass through.

ACTIVITY

What happens when light hits an opaque object?

A. Gather these materials: flashlight and sheets of red, green, blue, and white paper.

B. Darken the room. Hold 1 colored sheet of paper up to the white sheet. Shine the flashlight onto the colored paper.

   1. What color is reflected onto the white paper?

   2. What do you think happened to other colors that make up the white light of the flashlight?

C. Repeat step B for the other colored sheets of paper.

   3. What color was reflected onto the white sheet for the other 2 colors?

   4. What happens when light hits an opaque object?

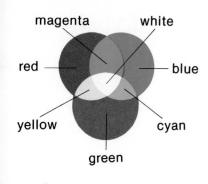

magenta white

red —   — blue

yellow   cyan

green

If you have ever mixed paints, you may know that the three primary colors of paints are red, yellow, and blue.

The color pictures in this book are made by the four-color printing process. Red, yellow, and blue inks are used. Black ink is the fourth color.

The primary colors of light are different from the primary colors of paints. Red, green, and blue are the primary colors of light. All colors can be made by mixing different amounts of these colors. As you can see in the diagram, red and green make yellow. Red and blue make magenta. What color do green and blue make? If the three primary light colors are mixed together, white is formed.

Section Review

Main Ideas: We see colors because of what happens when light hits an object. An opaque object like a green leaf reflects green and absorbs all the other colors. A transparent object, like a blue glass vase, absorbs all the colors but blue. The blue light passes through.

Questions: Answer in complete sentences.

1. When light shines on orange paper, what happens to the white light?
2. If you shine white light through a transparent red vase, what happens to the white light?
3. If red light is reflected from an object, what is the color of the object?
4. What are the primary light colors?

CHAPTER REVIEW

Science Words: Copy the numbered letters and spaces on paper. Use the hints to help you identify the science terms.

1. <u>m</u> __ __ __ __ __
2. __ __ __ <u>s</u>
3. __ __ <u>i</u> __ __
4. __ <u>p</u> __ __ __ __ __ __
5. __ __ __ __ __ <u>b</u>
6. __ __ <u>c</u> __ __ __ __ __ __ __
7. __ __ __ <u>e</u> __ __ __ __ __

Hints

1. Something you see that isn't really there
2. A curved piece of glass that bends light
3. A piece of glass that can form the spectrum
4. The seven colors of white light
5. To take in and hold light
6. An instrument that makes small things look larger
7. An instrument that makes faraway things seem close

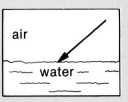

Questions: Answer in complete sentences.

1. What will be the path of light in the diagram? Draw a diagram to show your answer.
2. Light shines on the prism. What do you think will happen to the white light? Make a diagram to show your answer. You may use crayons.

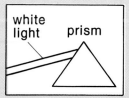

109

What happens when you mix the primary colors of light?

In this chapter, you learned that white light is made up of three primary colors: red, green, and blue. You will now see how you can mix these colors to produce both white light and other colors of light.

A. Gather the following materials: squares of red, green, and blue cellophane; 3 flashlights (or penlights); 3 rubber bands; and white cardboard screen.

B. Cover 1 flashlight with red cellophane. Fold over the edges and hold them in place with a rubber band. Repeat this step with the green and blue cellophane.

C. Darken the room. Shine colored spots from each flashlight on a white screen. Make sure all 3 flashlights are the same distance from the screen.

1. What color is produced when all three colors—red, green, and blue—shine on the same spot?

D. Try different combinations of the red, green, and blue lights to produce different colors. Copy the chart below and use it to record your observations.

Colors Mixed	Color Produced
1. red + blue	
2. red + green	
3.	
4.	

Photographer ▶

In 1826, a Frenchman made the first photograph.

A career in photography requires that you know about light and how lenses work. You also must know when to use different kinds of film and how to develop film. Photographers can work for newspapers, public relations companies, advertising firms, and many other businesses.

◀ Optician

The person shown in the picture is an optician. An optician makes the lenses in eyeglasses. The lens in the eyeglass helps the lens in your eye work properly. Together, they bend the light beams so that the beams focus exactly on the back of your eye. When the light beams focus this way, you see a clear picture. Opticians are skilled persons.

111

TOMORROW'S WEATHER

UNIT 3

CHAPTER 7

WEATHER

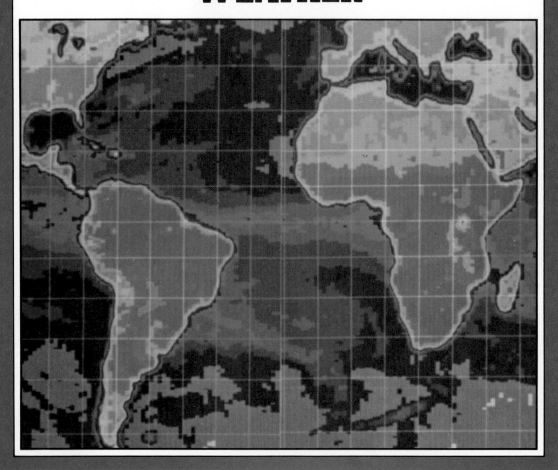

7-1.

The Ocean of Air

Pretend you are orbiting the earth. You can see a blue haze around the planet. This haze is a layer of air. No living thing can survive without it. When you finish this section, you should be able to:

114

□ **A.** Describe and compare the layers of the *atmosphere*.

□ **B.** Explain how to measure the temperature of the *atmosphere*.

□ **C.** Explain what causes the *atmosphere* to heat up in some places more than in others.

The earth's **atmosphere** (at-muh-sfeer) is made of air. Air contains many gases that you cannot see. There are two main gases in the air. They are nitrogen and oxygen. Without oxygen, you would not be able to breathe. The pie graph shows how much of each gas is in the air. The air also contains two other gases. They are water vapor and carbon dioxide.

Atmosphere: The air that surrounds the earth.

The *atmosphere* is made of at least four layers. The diagram on page 116 shows the four main layers of earth's atmosphere. The air is composed of tiny **molecules** (mahl-uh-kewls) of gas. The *molecules* in the bottom layer are tightly packed. Higher up, the molecules are spread farther apart. Let's take a closer look at each layer.

Molecule: The smallest particle of a substance.

The first layer is the **troposphere** (trope-uh-sfeer). This is the layer of air you breathe. The *troposphere* is between 8 and 16 kilometers (5 and 10 miles) thick. Most of the clouds you see are formed in the troposphere. The molecules of air in the troposphere at sea level are very close together. But on top of a mountain, the molecules are spread farther apart. In order to get

Troposphere: The layer of air closest to the earth (from 8 to 16 kilometers).

enough oxygen, you would have to breathe more deeply.

Above this first layer is the **stratosphere** (**strat**-uh-sfeer). It goes up to about 48 kilometers (30 miles) above the earth's surface. Most jet planes fly in the *stratosphere*. There is not much weather here, so flights are smoother. The molecules in the stratosphere are far apart. Because of this, jet planes carry their own supply of oxygen. The air in the plane is under pressure. This means that the molecules are tightly packed.

Above the stratosphere is the **mesosphere** (**mez**-uh-sfeer). The molecules are spread very far apart at this height. The *mesosphere* reaches to about 80 kilometers (50 miles) above the earth.

Stratosphere: The layer of air from 16 to 48 kilometers above the earth.

Mesosphere: The layer of air from 50 to 80 kilometers above the earth.

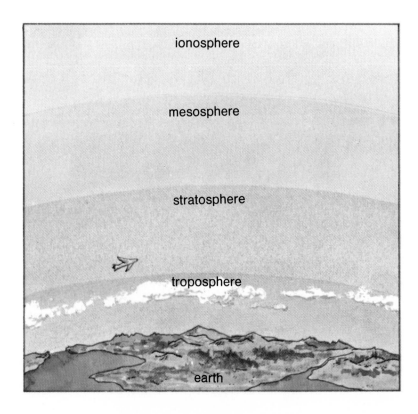

ionosphere

mesosphere

stratosphere

troposphere

earth

The last layer of the atmosphere is the **iono-sphere** (eye-**ahn**-uh-sfeer). It goes to about 1,000 kilometers (620 miles) above the earth's surface. The air is thinnest in the *ionosphere*. The molecules are spread very far apart. Where does the earth's atmosphere end and outer space begin? Why is it hard to know for sure?

The energy to warm earth's air comes from the sun. Sunlight can pass through the earth's atmosphere. The air does not heat up as the sunlight passes through it. Instead, the earth's surface heats up the air.

Sunlight passes through the air. It strikes the surface of the earth. The air over a warm patch of ground heats up. Have you ever noticed that some surfaces get hotter than others? Dark pavement feels very warm if you walk on it in bare feet on a sunny day. The air over the dark pavement will heat up faster than the air over grass or trees.

The **temperature** (**tem**-per-a-chure) of air gets cooler as you go higher up in the atmosphere. This is because the air is farther from the warm surface. Also, the molecules are spread farther apart. Molecules that are far apart hold less heat than molecules that are close together.

The *temperature* of the air is measured with a thermometer. When the temperature is warm, the liquid in a thermometer expands. It rises up the thermometer tube. When the temperature is cool, the liquid shrinks inside the tube. The level of the liquid falls.

Ionosphere: The layer of air farthest from the earth (up to 1,000 kilometers).

Temperature: The degree of hotness or coldness of the air.

ACTIVITY

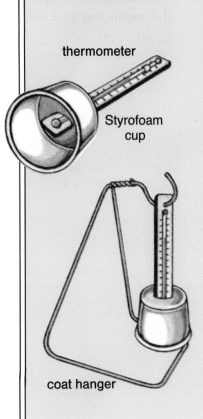

thermometer

Styrofoam cup

coat hanger

How does the surface of the earth affect air temperature?

A. Gather these materials: coat hanger, thermometer, and Styrofoam cup.
B. Bend the coat hanger, as shown. Push the thermometer through the Styrofoam cup. Hang the thermometer on the coat hanger.
C. Go outside. Measure the temperature of the air: (a) over blacktop, (b) over concrete, (c) in the shade of a tree, and (d) in the shade of a building.
   1. Was the temperature higher in direct sunlight or in the shade?
   2. How does the temperature differ over dark and light surfaces?
   3. Why is the air warm in some places and cool in others?

Section Review

Main Ideas: The earth is covered with a blanket of air called the atmosphere. It has four layers.

Questions: Answer in complete sentences.
1. What are the two main gases in the air?
2. In which layer are most of the earth's clouds found?
3. Why does the atmosphere heat up more in some places than in others?

118

Pretend you are hiking along a mountain ridge. It's been sunny and warm all day. But you see a dark line of clouds moving from the west. Do you think this means a storm is coming? It might help to know what kind of clouds they are. When you finish this section, you should be able to:

Observing the Weather

☐ **A.** Identify the elements that make up the earth's weather.

☐ **B.** Compare the three kinds of clouds.

☐ **C.** Identify the kind of weather related to each cloud type.

The following is a daily weather report:

It will be partly cloudy and hot in the city on Friday, with a 30-percent chance of an afternoon thundershower.

Why is it important for people to know what the weather will be like?

Weather is the condition of the air around the earth. Clouds, rain, temperature, and wind are all parts of weather. Scientists who study the weather are called **meteorologists** (mee-tee-or-**ahl**-uh-jists). *Meteorologists* help tell what kind of weather is coming.

The three main types of clouds are shown in the pictures on this page and the next. Symbols are used for each cloud type. They are shown in the lower right-hand corner of the picture.

Picture 1 shows clouds that look like feathers or curls. These clouds are found high in the sky. They are called **cirrus** (**seer**-us) clouds.

Meteorologist: A scientist who studies the weather.

Cirrus clouds: Clouds found high in the sky that look like feathers or curls.

120

Picture 2 shows low, flat sheets of gray clouds. These clouds spread out over the sky. They are called **stratus** (**strat**-us) clouds. *Stratus* clouds turn into rain clouds as they become bigger and thicker.

Picture 3 shows patches of puffy, white clouds. They look like cotton. They are called **cumulus** (**kyoom**-yoo-lus) clouds. *Cumulus* clouds are fair-weather clouds. They also become rain clouds as they get bigger and thicker.

Rain is only one kind of **precipitation** (prih-sip-uh-**tay**-shun). Moisture can also fall in the form of snow, sleet, and hail. The kind of *precipitation* that falls depends on the temperature of the air.

Stratus clouds: Low, flat sheets of gray clouds that spread out over the sky.

Cumulus clouds: Patches of puffy, white clouds.

Precipitation: Moisture that falls from the sky.

ACTIVITY

How are weather observations made?

A. Gather this material: thermometer.

B. Copy the weather chart shown below. Complete the chart for each day of the week.

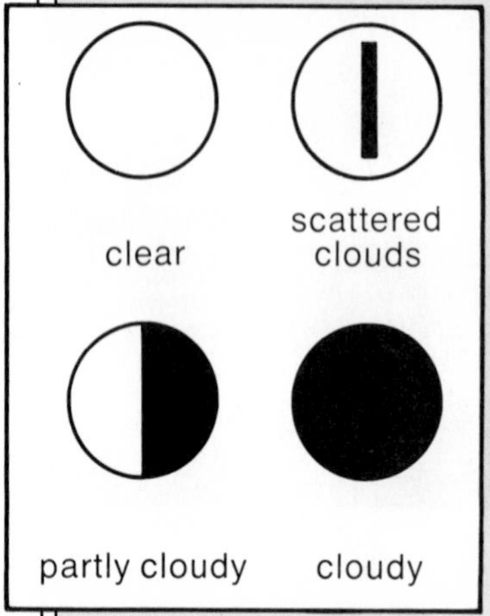

WEEK OF—				
	Day 1	**Day 2**	**Day 3**	**Day 4**
Cloud type				
Cloud cover				
Temperature				

C. Record the symbols for today's cloud type and cloud cover.

D. Measure the air temperature.

E. Repeat steps C–D each day for a week.
   1. What kind of clouds did you see most often?
   2. What was the weather like?

clear

scattered clouds

partly cloudy

cloudy

Section Review

Main Ideas: Cirrus, stratus, and cumulus are three types of clouds.

Questions: Answer in complete sentences.

1. What elements make up the earth's weather?
2. What kind of cloud is described by each of the following: (a) high, feathery clouds; (b) white, puffy clouds; (c) flat sheets of gray clouds?

Weather Forecasting

Some people think groundhogs can forecast the weather. February 2 is Groundhog Day. On this day, the groundhog comes out of its hole in the ground. If the groundhog does not see its shadow, winter is over. If the groundhog sees its shadow, it jumps back into its hole. Winter will last six more weeks. By using this method, you would be right about 30 percent of the time. What are some other strange ways to forecast the weather? When you finish this section, you should be able to:

■ **A.** Explain how information about the weather is recorded.

■ **B.** Explain how weather forecasts are made.

■ **C.** Explain how computers help to make forecasts.

Meteorologists collect information about the weather in order to make weather forecasts. The information includes temperature, air pressure,

and humidity. There are over 9,000 weather stations around the world. Weather stations collect a great deal of information. The weather forecaster's job is to use this information to make a good forecast.

A weather forecast is like a guess. A forecaster might say, "There is a 70-percent chance of rain tomorrow." The forecaster has studied many weather conditions. Wind direction, cloud types, temperature, and air pressure are all studied. Past weather conditions are helpful in making forecasts. For example, weather records show that it rained seven out of ten days with similar conditions. The forecast will be a 70-percent chance of rain.

The modern weather forecaster must use a computer. Without the help of a computer, it would be impossible to deal with all the information. Computers can be used to make maps. They also keep track of records to make forecasts. Computers can record weather information over many years. Weather scientists use this

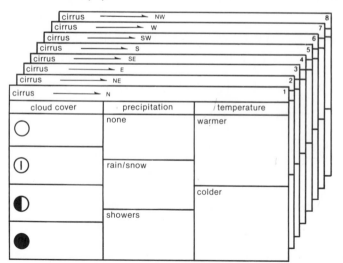

information to predict general trends in our weather.

You can make weather cards like those on page 124 to predict the weather. Label each card with a different cloud type and wind direction. Suppose today there are cirrus clouds. The wind is from the southwest. Look for that card. On the following day, record the weather on that card. Each time there are cirrus clouds and the wind is southwest, do the same. Do this ten times. Eight of these times it rained the next day. Twice it did not. The next time there are cirrus clouds and the wind is southwest, you can make a prediction. The chances that it will rain the next day are 8 out of 10, or 80 percent.

ACTIVITY

How is weather predicted?

A. Gather these materials: thermometer and weather cards.

B. Find the weather card for the kind of clouds and wind direction you had yesterday. Make a mark in the column for the kind of clouds you see today. Make a mark in the column for the kind of precipitation you are having today.

C. Measure the air temperature in the shade. Record whether today is warmer or colder than yesterday.

D. Repeat steps B–C each day for a week.

Section Review

stratus	SW	18
cloud cover	precipitation	temperature
◯ ǀ	none ЖЖ ЖШ	warmer ЖЖ ǁ
◑ ЖЖ ǁ	rain/snow ǀ	
◐		colder ǁ
● ǀ	showers	

Main Ideas: Weather records can be used to forecast the weather.

Questions: Answer in complete sentences.

1. The temperature is 13°C (55°F). It is raining. The wind is from the southwest. The clouds are stratus. Look at the card shown here. What are the chances for rain the next day?
2. What kind of information is used to make a weather forecast?
3. What are two ways in which computers help scientists forecast the weather?

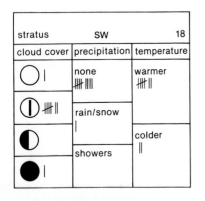

Vilhelm Bjerknes

Vilhelm Bjerknes (**byerk**-nes) was born in Oslo, Norway, in 1862. He devoted his life to the study of the weather. His dream was to be able to predict the weather just one day in advance. Weather forecasters today make predictions weeks and months ahead. Vilhelm would have been happy to make a 24-hour forecast. Around 1917, he set up a number of weather stations. These stations were located along the coast of Norway. Vilhelm studied the weather reports from all these stations. From these reports, he hoped to forecast the weather. He formed a team of scientists who began the science of meteorology.

CHAPTER REVIEW

Science Words: Think of a word for each blank. List the letters **a** through **l** on paper. Write the word next to each letter.

The air that circles the earth is the ___**a**___. Air is made of small particles called ___**b**___. People live in the part of the air called the ___**c**___. The air above 16 kilometers is called the ___**d**___. Above this layer is the ___**e**___. The top layer of the air is the ___**f**___.

Persons who study the weather are called ___**g**___. They use a thermometer to observe the air ___**h**___. High clouds are called ___**i**___. Low, flat sheets of gray clouds are called ___**j**___ clouds. ___**k**___ clouds are fair-weather clouds. Rain is one form of ___**l**___.

Questions: Answer in complete sentences.
1. What gases make up the earth's atmosphere?
2. How does the sun heat the earth's air?
3. Where would the temperature be higher on a sunny day: (a) above a black asphalt road; (b) above a light wooden picnic table; or (c) in the shade of a tree?
4. What are three things that meteorologists observe in making a weather report?
5. Which type of clouds would you expect to find on the following days: (a) a rainy day; (b) a partly cloudy day; and (c) a fair-weather day?
6. Are weather forecasts right all the time? Explain your answer.
7. Why are computers helpful in making weather forecasts?

CHAPTER 8

AIR ON THE MOVE

8-1.

Air Rises and Falls

The hot-air balloon was the first flying machine. Balloons like this one can rise thousands of feet into the air. Some rise so high that the crew has to take their own oxygen. Very soon, the balloon will rise off the ground. What causes the balloon to rise? When you finish this section, you should be able to:

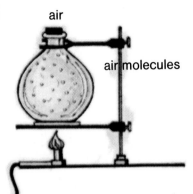

air

air molecules

- **A.** Explain what happens when air is heated or cooled.

- **B.** Explain how the temperature of air affects the *air pressure*.

Air is made up of particles called molecules. Picture each molecule of air as a tiny, round ball. If you heat these tiny molecules, they start to move. The more you heat them, the faster they move. If you cool them, they slow down. Suppose you put popcorn kernels in a hot pan. The kernels will start to move. They bounce around the pan. As the pan cools, the kernels stop moving. Air molecules move in the same way.

corn kernels

We can now find out why heating the air in a balloon makes it rise. Look at the drawings below. Before they are heated, the air molecules inside the balloon are tightly packed. When the

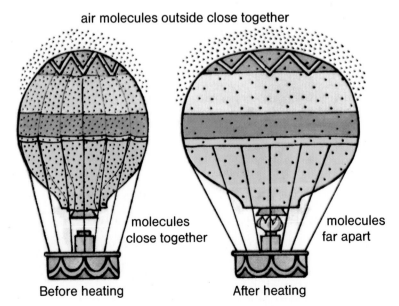

air molecules outside close together

molecules close together

molecules far apart

Before heating

After heating

129

Air pressure: The weight of the atmosphere caused by molecules pushing down on the earth's surface.

Barometer: An instrument used to measure the air pressure.

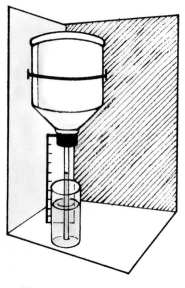

air is heated, the molecules move farther apart and the volume of air increases. The molecules push on the inside of the balloon. This makes the balloon expand. Since the molecules are farther apart, the air inside the balloon is lighter than the air outside the balloon. The lighter balloon will rise into the air.

Suppose you could weigh two columns of air as shown in the drawing. One column is cold air. The other is warm air. Would one column be heavier? Why? Air molecules in cold air are close together. There are more molecules in the cold-air column. So the cold-air column weighs more and pushes down with more force or pressure on the balance.

The force with which the air pushes down is called **air pressure**. *Air pressure* is measured with a **barometer** (buh-**rahm**-uh-ter). In many *barometers*, a tube is sealed at one end. The open end of the tube is placed into a jar. The jar is filled with a liquid. Air presses down on the liquid. The liquid is pushed up the tube. The greater the air pressure, the higher the liquid rises in the tube. When the air pressure goes down, the liquid moves down the tube.

Weather reports include the air pressure. Air pressure can change from day to day. Weather reports say if the pressure is rising, falling, or staying steady. You can predict the weather if you know how the air pressure is changing. The pressure may remain steady. If it does, the weather will not change very much. The chart on page 132 shows these forecasts.

ACTIVITY

How is the volume of air increased?

A. Gather these materials: broomstick, chair, 2 large paper bags, meter stick, paper clip, string, 150-watt bulb, and clay.

B. Make a balance like the one shown in the diagram. Bend the paper clip into a hook so the meter stick does not rub against the broomstick.

C. Hang the bags so the meter stick is level, or balanced. You can add a small amount of clay to the meter stick to make it balance.

D. Turn on the bulb. Hold it just inside 1 of the bags. Hold it there until you see a change.
  1. What change did you see?
  2. What do you think caused this change?
  3. What happens when air is heated?

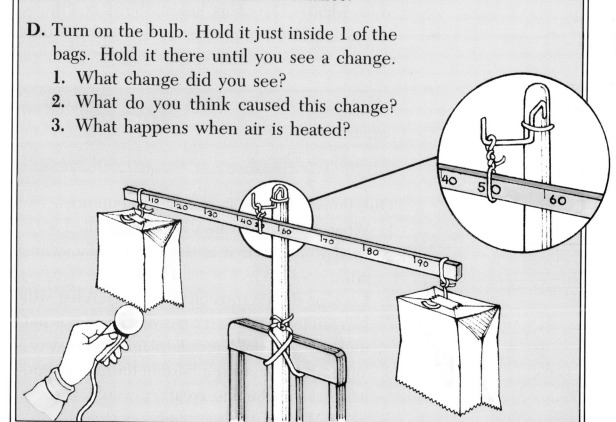

AIR PRESSURE PREDICTION CHART	
Air Pressure	**Change in the Weather**
Rising ↑	Clear, no rain
Falling ↓	Cloudy with rain
Steady	Little change

When you listen to a weather forecast, you will hear about high- and low-pressure centers. High pressure brings sunshine and clear weather. Low pressure means some form of precipitation, such as rain or snow.

Section Review

Main Ideas: The chart below shows how temperature affects weight, pressure, and the distance between air molecules.

	Weight	**Pressure**	**Molecules**
Warm Air	light	low	far apart
Cold Air	heavy	high	close together

Questions: Answer in complete sentences.
1. What will happen to a beach ball filled with air if you release it at the bottom of a swimming pool? Why?
2. Look at the picture on page 131. What will happen to a balance if a pan of ice cubes is put under one of the bags? Explain your answer.
3. When the liquid in a barometer goes down, what change in the weather would you expect? What if the barometer is steady?

Have you ever tried to fly a kite on a day when the wind was not blowing? Without wind, the kite will not fly. There are some places where a steady wind always blows. The child shown here is flying his kite at the beach. The sun is shining. Is this a good place to fly a kite? When you finish this section, you should be able to:

- ☐ **A.** Explain how changes in air pressure and temperature cause wind.

- ☐ **B.** Explain why the direction in which the wind blows at the seashore changes.

If you put an object in the path of sunlight, the object will heat up. Dark-colored objects heat up more than light-colored objects. The earth is heated by sunlight. Some places get warmer than others. For example, dark surfaces can get very hot. They absorb a lot of sunlight. Light surfaces, like water, reflect a lot of sunlight. As a result, water does not get hot very fast. It takes much longer to heat water than land. But dark objects also cool faster. At night, the land will cool faster than the water.

Heat from the earth warms the air. In the same way, heat from a radiator warms the air in a room. As the earth warms and cools, the air near the earth's surface also warms and cools. On a warm day at the beach, the air over the land is warmer than the air over the water. The daytime air heated by the land floats upward. The colder air over the water moves in under the warmer, rising air. This happens because cold air pushes with more pressure than warm air. Air always moves from places of high pressure to places of low pressure. The moving air is called a breeze or a wind. What do you think happens at night, when the land cools faster than the water?

ACTIVITY

Which heats faster: soil or water?

A. Gather these materials: 2 bowls, potting soil, water, 2 thermometers, and watch.

B. Put water in 1 bowl. Put potting soil in the second bowl. Fill both bowls to the same depth.

C. Place thermometers in both bowls. Put both bowls in sunlight.

TEMPERATURE					
	Start	**3 min**	**6 min**	**9 min**	**12 min**
Soil					
Water					

D. Copy the chart shown here. Record the temperatures of the water and soil after 3 min, 6 min, 9 min, and 12 min.

1. Did the temperature of the soil change? How much?
2. Did the temperature of the water change? How much?
3. Which heats faster: soil or water?
4. Why do you think there is a difference in the way soil and water heat up?

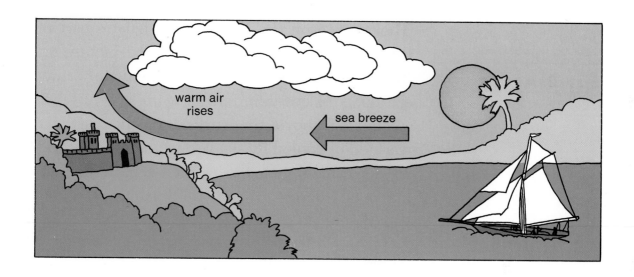

In the diagram above, the pressure over the water is higher than the pressure over the land. The greater the difference in pressure between two places, the faster the wind will blow.

Section Review

Main Ideas: As the earth warms and cools, the air near the earth's surface also warms and cools. Air always moves from places of high pressure to places of low pressure. Moving air is called wind.

Questions: Answer in complete sentences.
1. Which of these surfaces do you think will heat up the fastest: (a) gray car; (b) grassy field; (c) tar road? Explain your choice.
2. Is the kite shown flying in the wrong direction? If so, show the right direction.
3. In the picture, what is the direction of the wind?
4. Is the temperature higher at A or B?

135

8-3.

Air Masses

The picture on the left shows a barren part of the Arctic near the North Pole. The picture on the right shows a desert near the equator. How are these places different?

The air above the surface of the earth is warmed or cooled by the surface below it. Is the air hot or cool over the Arctic? What is the air like over a desert? When you finish this section, you should be able to:

☐ **A.** Describe what *air masses* are and where they form.

☐ **B.** Compare *air masses* in terms of temperature and moisture.

☐ **C.** Explain what happens when *air masses* meet.

If air were to remain near the North Pole, it would get very cold. Air staying near the equator would get very warm. An **air mass** is a large amount of air. *Air masses* can be hot or cold. What kind of air masses would you expect to form near the poles? What kind would form near the equator?

Air masses also can be wet or dry. Deserts are hot, dry places. What would the air mass over a desert be like?

Look at the map above. Six air masses push into each other over North America. There are three blue air masses. These are all cold air masses. There are three orange air masses. These are hot air masses. Some air masses form over land. Some form over water.

The two air masses that form over the land are dry. Dry air mass E is high-pressure air. This air

Air mass: A large volume of air that takes on the temperature and moisture of the area over which it forms.

mass forms over deserts. It pushes to the north and east. It brings clear, hot, and dry weather. Dry air mass B is high-pressure air. It forms in Canada. It pushes to the south and east. It carries clear, dry, and cold weather to the United States.

The four air masses (A, C, D, F) over the oceans pick up moisture. Air masses A and C bring cold, high-pressure moist air. Air masses D and F bring warm, low-pressure, moist air. The moisture the air masses pick up falls to earth as precipitation.

Places covered by the same air mass have the same kind of weather. The temperature of these places is about the same. The amount of moisture in the air is about the same. Storms occur where air masses meet. The picture below was taken by a camera in space. Can you tell if it was sunny or cloudy where you live?

ACTIVITY

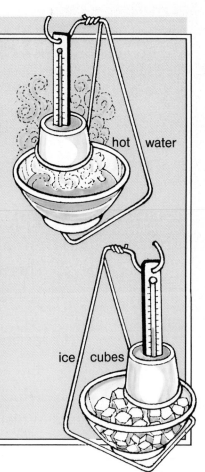

How is an air mass formed?

A. Gather these materials: 2 bowls, hot water, ice, 2 thermometers, 2 coat hangers, and 2 plastic cups.

B. Fill 1 bowl with ice. Fill the other bowl with hot water.

C. Slip inverted cups over lower end of thermometers and hang over bowls.

D. Measure the temperature over each bowl each minute for 5 min.

  1. Was the temperature different over the 2 bowls at the end of 5 min? Why?

  2. Compare the kind of air masses that would form over the Arctic Ocean and an ocean near the equator.

hot water

ice cubes

Section Review

Main Ideas: Air masses form when air remains over a cold or hot place. Six large air masses affect our weather.

Questions: Answer in complete sentences.

1. What is an air mass?
2. What kind of air mass would form over an ocean near the South Pole?
3. Where do you think a hot, moist air mass would form?
4. What type of weather would you expect where air masses meet?

CHAPTER REVIEW

Science Words: Use the clues to fill in the words in the spaces below.

1. _ _ _ _ _ _ _ _ _
2. _ _ _ _ _ _ _ _ _ _ _
3. _ _ _ _ _ _ _

Clues:

1. An instrument used to measure air pressure
2. The weight of the air (two words)
3. A large amount of air (two words)

Questions: Answer in complete sentences.

1. The scale is weighing columns of hot and cold air. Which way will the balance tip? Why?
2. Which column of air is pressing down more?
3. How could you stop a hot-air balloon from rising?
4. What does a barometer measure?
5. During the day, will the wind blow (a) from land to ocean; or (b) from ocean to land?
6. What happens to the air above the ground when the sun shines?
7. What kind of air masses will form over these places: (a) central Canada near the Arctic Circle; and (b) the ocean near the equator?

8. On the map to the left, show where you think these air masses would form: (a) cold, moist air; (b) hot, dry air; (c) hot, moist air; and (d) dry, cold air.
9. Make an air pressure chart showing what kind of weather you would expect if a barometer (a) remains steady, (b) rises, and (c) falls.

CHAPTER 9

WATER IN THE AIR

It often rains when we are not prepared for it. Dark clouds form and rain begins to fall.

Where does rainwater come from? What causes it to rain? When you finish this section, you should be able to:

9-1.

Rain and Clouds

A. Explain what the *dew point* is.

B. Explain how *water vapor* gets into the air.

C. Describe how water moves through the *water cycle*.

Water moves back and forth between the air and the ground. Suppose you place a saucer of water on a windowsill. Later, you see the water is gone. Why? When the water dried up in the saucer, it changed to a gas. This change from a liquid to a gas is called **evaporation** (ee-vap-uh-**ray**-shun). The gas goes into the air. Water in the form of a gas is called **water vapor**.

Water vapor in the air can change back into a liquid. This happens if the air is cooled. The water vapor will condense on a surface such as a blade of grass. **Condensation** (kahn-den-**say**-shun) happens when water vapor changes to liquid water. The temperature at which water vapor *condenses* is called the **dew point**. When the temperature of the air falls to the *dew point*, water forms on the grass.

Evaporation: The change of a liquid into a gas.

Water vapor: Water in the form of a gas.

Condensation: The change of a gas into a liquid.

Dew point: The air temperature at which water condenses.

Water vapor needs an object to condense on. Clouds form when water vapor condenses on dust in the air. Without dust, there would be no clouds.

As more water vapor condenses, the drops of water become larger and heavier. They fall to the earth as precipitation. The symbols in the margin are used for the kinds of precipitation. Use them on your daily weather chart.

What do you think happens to the water that falls as precipitation? First, it evaporates. Next, it condenses into clouds. Then it falls to the ground as precipitation again. The evaporation and condensation of water is called the **water cycle** (sy-kul). A *cycle* describes something that happens over and over again in the same order. The drawing below shows these steps in the water cycle.

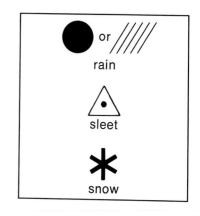

rain

sleet

snow

Water cycle: The evaporation and condensation of water over and over again.

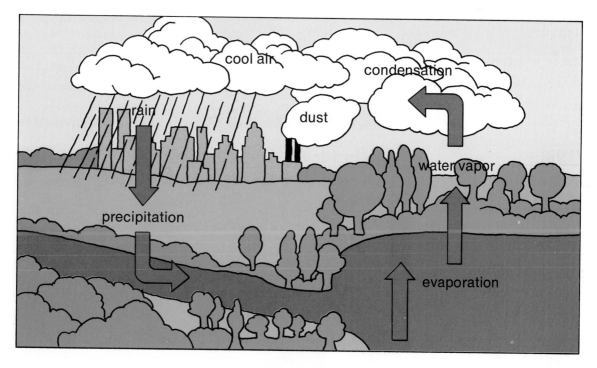

ACTIVITY

What happens to water vapor when the air is cooled?

A. Gather these materials: ice cubes, metal can, thermometer, and water.

B. Place the ice cubes in a metal can half filled with water. Put the thermometer in the can. Stir the ice slowly with the thermometer. Watch the outside of the can. When water forms on the outside, read the temperature.

C. Repeat step B outside.

1. What was the temperature of the ice water when dew formed on the can inside?

2. What was the temperature of the ice water when this happened outside?

3. What was the dew point inside and outside?

Section Review

Main Ideas: Water moves back and forth between the air and the surface of the earth.

Questions: Answer in complete sentences.

1. What is the difference between evaporation and condensation?

2. If air is cooled enough, what happens to the water vapor in it?

3. Describe the water cycle.

Cold and Warm Fronts

Pretend you are the pilot of an airplane. You are flying into the clouds shown here. You will be going from one air mass to another. Do you think this change will affect the people on your plane? When you finish this section, you should be able to:

☐ **A.** Describe how the weather changes when a cold *front* moves across an area.

☐ **B.** Describe how the weather changes when a warm *front* moves across an area.

You learned that cold and warm air masses move away from the place over which they form. The air masses move toward each other. They meet in an area called the zone of mixing. The small map shows the zone of mixing. Do you live in this zone?

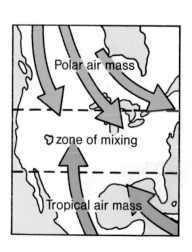

145

ACTIVITY

What happens when a cold air mass meets a warm air mass?

You will make a model of the air using water. Models let you see a part of nature in the classroom.

A. Gather these materials: plastic shoe box, piece of plastic or cardboard, water, red food coloring, and salt.

B. Fill the box with water. Using a piece of plastic, divide the box into 2 sides.

C. Quickly pour salt into one half of the tank. Stir gently. Add red food coloring to this side. The salt water is heavier. It is a model for cold air. The water in the other side is lighter. It is like warm air.

1. What do you think will happen when you lift the plastic between the 2 sides?

D. Lift the plastic.

2. What did you see when the 2 kinds of water met?

3. How is this like a front? What kind of front?

Jet stream: Fast-moving winds high in the troposphere.

High in the troposphere is a fast-moving stream of air. This is called the **jet stream**. It moves from west to east. It goes through the zone of mixing. The *jet stream* helps to mix warm and cold air masses. It also helps to move them from west to east.

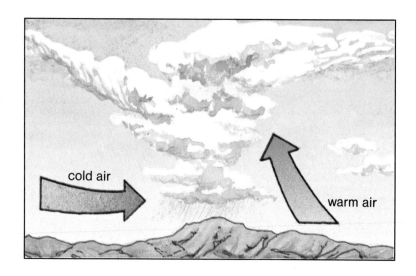

cold air

warm air

The places where cold and warm air masses meet are called **fronts**. If cold air is pushing into warm air, it is a cold *front*. If the warm air is doing the pushing, it is a warm front.

Look at the drawing of the cold front. The front edge of the cold air mass is steep. It pushes the warm air up off the ground quickly. The rising warm air cools. Clouds form. Cold fronts often bring showers. There may be thunder and lightning with the shower. The air behind a cold front is clear and cooler. Fair-weather cumulus clouds are carried by the clear, cooler air.

The drawing on page 148 shows that a warm front is long and sloping. The warm air rides up over the cold air and pushes it away. As a warm front moves in, many kinds of clouds form. High feathery cirrus clouds are seen first. In the next 12 to 24 hours, the clouds thicken. They drop lower in the sky. They change to stratus clouds. Then they change to rain clouds. Warm fronts almost always bring rain or snow.

Fronts: The places where cold and warm air masses meet.

147

warm air

cold air

Very small drops of slowly falling rain are called **drizzle** (**driz**-ul). Dark stratus clouds bringing *drizzle* may follow a warm front. Warm fronts may take up to two days to pass. Cold fronts usually pass in a few hours.

Sometimes warm fronts pass in less than two days. This happens if cumulus clouds follow cirrus clouds. You may have heavy rain within 12 hours. There may even be thunder and lightning.

Drizzle: Very small drops of slowly falling rain.

Section Review

Main Ideas: There are two types of fronts: cold and warm.

Questions: Answer in complete sentences.

1. Storms were followed by clear, cooler air. What kind of front passed?
2. The clouds changed from cirrus to stratus. What kind of front passed?
3. What type of front travels very fast?
4. The temperature is forecast to drop quickly. What type of front do you think will pass through?

The Harrison family is looking at a weather map. They are planning a trip from St. Louis to Denver. They want to know what the weather will be like as they drive west. The weather map will tell them what kind of weather to prepare for. When you finish this section, you should be able to:

☐ **A.** Identify the direction in which weather moves across the United States.

☐ **B.** Read a weather map.

☐ **C.** Forecast the weather a day or two ahead by using a weather map.

A weather map is a tool to a weather forecaster. Daily newspapers often print a weather map of the whole country. The map uses symbols to describe the weather. These symbols are shown on the next page. What symbol is used for a high pressure center? How can you tell a cold front from a warm front? How are thunder-

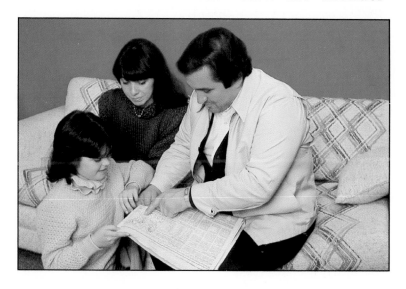

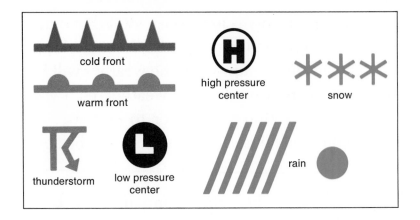

storms shown on the map? How is the symbol
for rain different from the symbol for snow?

The map below is a weather map for a Monday. Study this map. Then answer these questions. Over what state is the low-pressure center on the map? What do the red and blue lines with the bumps and points show? Suppose the map were printed in black and white. How could you tell the warm front from the cold front? The points on the cold front show the direction the cold front is moving. Cold air is behind the cold front. Where is the warm air? How do you know?

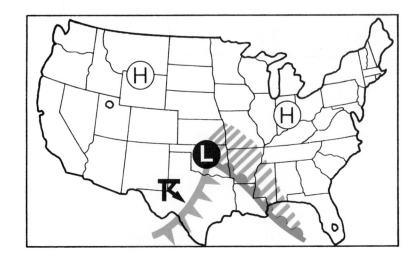

Now look at the map for Tuesday. You can see that the low-pressure center has moved. In which direction did it move? Did the high-pressure centers move in the same direction? There is rain ahead of the warm front. Most of the rain is near the low-pressure center. Low-pressure centers are usually the centers of storms. What kind of storms are along the cold front?

The map for Wednesday shows a new low-pressure center. The old one has moved farther east. Low-pressure centers are followed by high-pressure centers. The highs and lows travel mainly from west to east. Look at the warm front and the cold front. They have been on the map for three days. The cold front has almost caught up with the warm front. Cold fronts move faster than warm fronts.

All weather forecasters use pictures of the earth. These pictures are made by satellites in outer space. The patterns of the clouds help forecast the weather. The pictures show how patterns of clouds move.

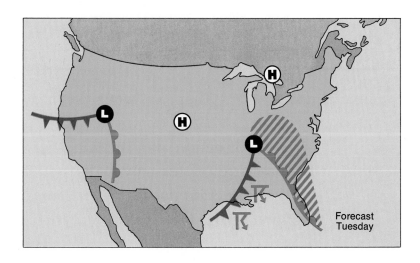

Forecast
Tuesday

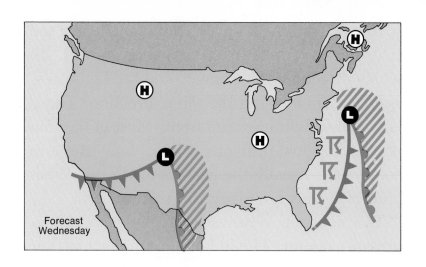

Forecast
Wednesday

ACTIVITY

How does weather move across the United States?

A. Gather these materials: outline map of the United States, weather maps from a newspaper for 3 days in a row, construction paper, glue, and scissors.

B. Glue the 3 U. S. weather maps on a piece of construction paper. Find a low or a high pressure on the first of the 3 maps. Mark the high or low on the outline map.

C. Find the same high or low pressure on the next 2 maps. Mark them on your outline map.

1. In what direction did the high- or low-pressure center move?
2. In what direction does weather seem to move across the United States?

Low-pressure centers show up as swirls of clouds. Forecasters can watch these swirls move. They compare pictures taken on different days. Computers show how fast and in what direction the weather is moving. The forecasters use this information to predict where the storm will move next.

Section Review

Main Ideas: Symbols on weather maps show kinds of weather. High- and low-pressure centers travel from west to east. Satellite pictures show how weather moves across the earth.

Questions: Answer in complete sentences.

For questions 1–3, look at the map on page 150.

1. Where do you think the low-pressure center above Oklahoma will move during the next day?
2. What kind of weather is Tennessee having?
3. What kind of weather do you think Tennessee will have on Thursday?
4. In what direction does weather move across the United States?

CHAPTER REVIEW

Science Words: Complete the crossword puzzle on a separate sheet of paper.

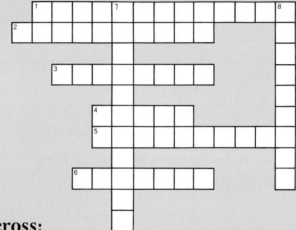

Across:

1. The change of a gas into a liquid
2. Water in the form of a gas (two words)
3. The air temperature at which water condenses (two words)
4. A place where cold and warm air masses meet
5. The evaporation and condensation of water over and over (two words)
6. Very small drops of slowly falling rain

Down:

7. The change of a liquid into a gas
8. Fast-moving winds high in the troposphere (two words)

Questions: Answer in complete sentences.

1. Draw a diagram of the water cycle. Label each part. Explain the cycle.
2. The weather report says a cold front will be moving into the city. What do you think the weather will be?

CHAPTER 10

WEATHER AND PEOPLE

Storms like tornadoes and hurricanes are violent. They can destroy homes. They cause flooding. People and wildlife may be harmed. How are tornadoes formed? When you finish this section, you should be able to:

Violent Storms

A. Identify four kinds of violent storms.

B. Compare the kind of weather each violent storm causes.

C. Describe the effects of violent storms on people.

Hurricane: A storm that forms over warm oceans near the equator.

Storms that form over the warm oceans near the equator are called **hurricanes** (her-uh-kaynz). *Hurricanes* are the strongest storms on earth. They bring heavy rains and very strong winds. They form between June and November. During these months, the sun is above the equator. It is so hot that large amounts of ocean water evaporate. The warm, moist air rises quickly. Then it condenses. A violent storm forms. The picture below shows what happens when a hurricane hits land.

Storms with thunder and lightning, heavy rains, and strong winds are called **thunderstorms** (**thun**-der-stormz). These storms often form when warm, moist air rises quickly. The water droplets in the clouds may become charged. Electric charges jump from cloud to cloud or from a cloud to the ground. We see lightning. The picture above shows lightning striking the ground during a thunderstorm.

Thunderstorms can cause small storms called **tornadoes** (tore-**nay**-dohz). A *tornado* is a spinning, funnel-shaped cloud. It touches the earth as it moves along. The picture on page 158 shows a tornado. Tornadoes are very violent storms. They can destroy things in their path.

Tornadoes occur in the spring from April to June. The weather is changing during this time of year. Many fronts move across the United

Thunderstorm: A storm that brings thunder, lightning, heavy rains, and strong winds.

Tornado: A storm that produces a spinning, funnel-shaped cloud.

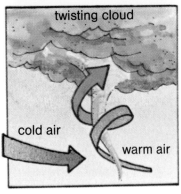

Blizzard: A storm that brings cold winds and blowing snow.

States. Tornadoes have occurred in all states in the country. But most of them occur in the South and Midwest.

Tornadoes form along cold fronts. Cold fronts move very fast. As the cold front pushes forward, it forces warm air to rise quickly. The rising air may twist. When this happens, a funnel-shaped cloud forms.

The winds inside a tornado can be as high as 800 kilometers (500 miles) per hour. As the tornado moves, it skips and jumps across the land. It is very hard to guess exactly where it will touch down. When a tornado touches down, it can destroy objects in its path.

The air pressure inside the tornado is very low. As the tornado comes close to buildings, a strange thing happens. If the windows and doors are closed, the air inside a building is trapped. As the tornado comes close, the air pressure outside the building gets very low. Inside the building, the pressure is high. The difference in pressure causes the air to rush from inside to outside. As a result, windows break and walls collapse.

Some storms bring cold winds and heavy, blowing snow. These storms are called **blizzards** (**bliz**-erdz). *Blizzards* are dangerous storms. The freezing winds blow at a speed of over 70 kilometers (43 miles) per hour. A blizzard can drop over 60 centimeters (2 feet) of snow. Roads become icy and slippery. Power lines may fall. People become stranded. The heating systems

in homes or other buildings may fail. People may freeze to death.

Section Review

Main Ideas: Hurricanes, thunderstorms, tornadoes, and blizzards are violent storms.

Questions: Answer in complete sentences.

1. What type of storm might form in each of these cases: (a) a fast-moving cold front moves into hot, humid air; (b) a front moves north, the temperatures are below freezing, and there are high winds; and (c) it is September, with high winds and heavy rain?
2. What effect could a tornado have on homes and buildings?
3. Why could buildings explode if a tornado passed?

10-2.

Hot, Wet Air

The inside of a greenhouse is very hot. The air is full of moisture. On some days, the outside air feels just like the inside of a greenhouse. When you finish this section, you should be able to:

☐ **A.** Explain what *humidity* is.

☐ **B.** Describe how the amount of *humidity* in the air affects people.

☐ **C.** Explain the effect of air temperature on *humidity*.

You learned that one path in the water cycle is evaporation. Water that falls to the ground as

rain goes back into the air. When water evaporates, it becomes water vapor. The air can hold only a certain amount of water vapor.

Water vapor in the air affects the way we feel. The amount of water vapor in the air is called **humidity** (hyoo-**mid**-ih-tee). On a hot summer day when the *humidity* is high, the air feels sticky. The air is holding a great deal of water. When the humidity is high, water condenses on the grass at night.

The weather service predicts hot, moist days using a *temperature-humidity index*. They call it *THI* for short. The higher the THI, the more uncomfortable you feel. Most people are uncomfortable if the THI is over 80. But if the THI is less than 70, most people feel comfortable.

On hot, humid days, the liquid you perspire is not easily evaporated. The air cannot hold much more moisture. This is part of the reason you feel hot and sticky. What effect does evaporation have on your body?

Humidity: The amount of water vapor in the air.

ACTIVITY

What happens to the temperature when water evaporates?

A. Gather this material: water.

B. Moisten the back of your hand. Wait a few seconds. Then blow across the wet spot.

  1. How did the wet spot feel?

C. Moisten the back of the same hand. Blow across the wet spot. Also, blow across the back of your other hand.

  2. How did the wet hand feel compared with the dry hand when you blew across them?

  3. What happened to the moisture when you blew across it?

  4. What happened to the temperature of your hand when you blew across the wet spot?

Hygrometer: An instrument used to measure the humidity in the air.

Meteorologists measure humidity with a **hygrometer** (hy-**grahm**-uh-ter). One type of *hygrometer* uses a strand of human hair. Human hair stretches when it is humid. A homemade hygrometer is shown on page 163. On a damp and cloudy day, the hair will stretch. The pointer will move. On a dry, sunny day, the hair will shrink. The pointer will move the other way.

A second type of hygrometer uses two thermometers. One of them is wet. The hygrometer is then whirled around. If the temperature on

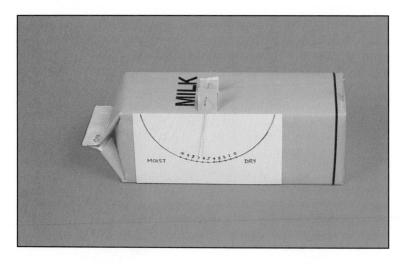

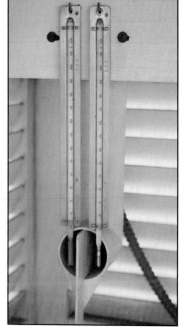

the wet bulb drops sharply, the humidity is very low. If the temperature does not drop much, the humidity is high.

Section Review

Main Ideas: The moisture in the air is called humidity. Humidity is measured with a device called a hygrometer.

Questions: Answer in complete sentences.

1. Why do you feel hot and sticky on humid days?
2. What happens to the temperature when water evaporates?
3. How does perspiring help you cool off on a hot day?
4. Why do you think dogs have to pant to stay cool?
5. Why does a room fan make you feel cool on a hot, humid day?
6. How can you measure humidity?

10-3.

The Changing Atmosphere

The city scene above shows how humans have affected the air in Los Angeles. Compare this picture to the one of mountain air. Why is the air so different in these two places? When you finish this section, you should be able to:

☐ **A.** Identify causes of *air pollution*.

☐ **B.** Explain how both human and natural causes can change the atmosphere.

☐ **C.** Identify some things we can do to reduce *air pollution*.

People need clean air to live. When we breathe, whatever is in the air enters our lungs. Some things in the air are harmful to us. People with breathing problems would be in danger if

they lived in a polluted area. **Air pollution** (pol-**oo**-shun) damages the lungs. It also hurts the eyes. It can even cause death.

People are not the only living things harmed by *air pollution*. Plants need sunlight in order to grow and remain healthy. If the air is polluted, less sunlight reaches plants' leaves. Also, dust and dirt in the air settle on the leaves. If too much dirt gets on the leaves, the plant could die from lack of sunlight.

Air pollution: The presence of dirt, dust, and chemicals in the air.

165

How can you see dust and dirt in the air?

A. Gather these materials: petroleum jelly, 4 plastic microscope slides, hand lens, and paper towels.

B. Put a small amount of petroleum jelly on each of the 4 slides. Use your finger. Any dust and dirt in the air will stick to the petroleum jelly.

C. Put slide no. 1 away in a safe, clean place.

D. Put the other slides in these places.

Slide no. 2: on the windowsill inside your classroom.

Slide no. 3: on the windowsill outside your classroom.

Slide no. 4: on the ground near a parking lot.

E. Leave the slides for 24 hours.

F. Collect all 4 slides. Compare each slide.

1. Which slide showed the most traces of dust and dirt?

2. What could be the source of dust and dirt in each place?

Air pollution has many causes. Humans are the major cause of air pollution. The pictures on page 167 show some of the ways humans pollute the air. What do you think is polluting the air in each picture?

Cars are a major cause of air pollution. Cars put many kinds of gases into the air. Air pollution is worse in large cities than in the country. Factories also pollute the air. Many of them burn coal. Coal-burning makes gases such as carbon dioxide. Too much carbon dioxide in the air causes air pollution.

Humans are not the only source of air pollution. Nature also pollutes the air. These two examples show how natural causes can pollute the air.

A volcanic eruption like the one shown can pollute the air. When a volcano erupts, it throws fine dust and gases into the air. The dust blocks some sunlight from reaching the earth's surface. The air temperature drops.

The atmosphere has also changed in the past.

Some scientists believe the earth was once hit by a giant **asteroid** (ass-ter-oyd). They think this took place a long time ago. When the *asteroid* hit, dirt and dust were thrown into the air. A dark cloud covered the whole earth. This cloud may have caused the temperature of the earth to drop. Many plants would have died because of the lower temperature. Since most **dinosaurs** (**dine**-uh-sorz) ate plants, this might explain why the *dinosaurs* died.

Clean air is important to life on earth. There are ways to keep the air clean. Cars can use special cleaners in their exhaust systems. This keeps unsafe gases and fumes from getting into the air. Factories can clean the smoke that goes out of their smokestacks by using special filters.

It costs a lot of money to reduce air pollution. Many people think it is too expensive. But in many ways, our lives depend on clean air.

Section Review

Main Ideas: Dust, dirt, and chemicals in the air are all air pollutants. Humans and natural events, like volcanoes, can pollute the air.

Questions: Answer in complete sentences.

1. What is air pollution?
2. How have humans polluted the air?
3. What effect does too much dust have on the atmosphere?
4. How can a volcano pollute the air?
5. How can air pollution be reduced?

CHAPTER REVIEW

Science Words: Find the words that are hidden in the puzzle below. Write the meaning of each word that you find.

```
H U R R I C A N E O P L V S O P H
U D I T H U N D E R S T O R M P Y
M I O T O R N A D O K L M Y U O G
I N K L I O B L I Z Z A R D I O R
D N A S T E R O I D D E R T U O O
I O O P O L L U T I O N J Y T U T
```

Questions: Answer in complete sentences.

1. What type of weather may occur with these storms: (a) tornado; (b) hurricane; (c) thunderstorm; and (d) blizzard?
2. What happens to the air temperature when water evaporates?
3. An air conditioner removes moisture from the air. Why does this make you feel cooler?
4. What is humidity?
5. What are some causes of air pollution?
6. In what way does a volcano pollute the air?
7. How can air pollution harm people?
8. Would the air be safer to breathe while jogging in the city or in the mountains? Explain.
9. How can air pollution from cars and factories be reduced?

INVESTIGATING

How can you use a barometer to forecast the weather?

You can make your own barometer to measure the air pressure. You will then be able to forecast the weather.

A. Gather these materials: 2 olive or baby food jars, balloon, ice cream stick, glue, plastic straw, 2 rubber bands, and scissors.

B. Cut a large section from the balloon. Stretch it tightly over 1 jar. Fasten it with a rubber band.

C. Cut a point on the straw. Glue the other end to the center of the balloon.

D. Fasten the stick to the other jar with a rubber band. Put the point of the straw next to the stick.

E. For the next week, call the weather bureau each day. Record the air pressure on the scale.

F. Put your barometer in a place where the temperature does not change. Do not move it once you have set it up.

  1. How does the air pressure change when the weather changes?

  2. How can you use air pressure to forecast the weather?

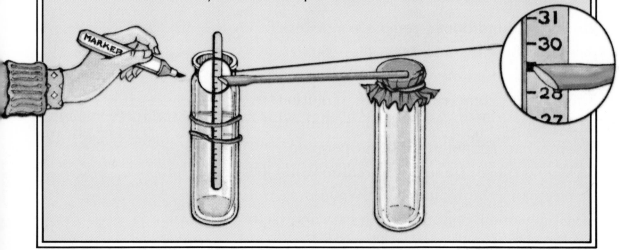

CAREERS

Television Weather Forecaster ▶

Each day people give weather reports and forecasts on television. Meteorologists use computers, instruments at the TV station, radar, and satellite pictures. All this information helps them predict the weather. Weather forecasters have studied meteorology in college.

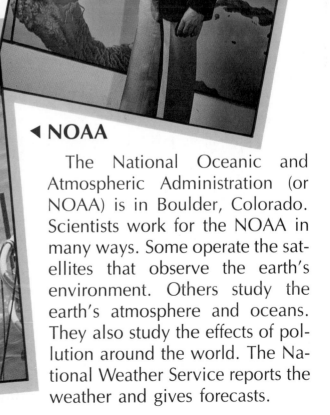

◀ NOAA

The National Oceanic and Atmospheric Administration (or NOAA) is in Boulder, Colorado. Scientists work for the NOAA in many ways. Some operate the satellites that observe the earth's environment. Others study the earth's atmosphere and oceans. They also study the effects of pollution around the world. The National Weather Service reports the weather and gives forecasts.

MACHINES

CHAPTER 11

MACHINES AND WORK

11-1.

What Is Work?

Many state fairs have a competition between teams of horses. The team that can move the heaviest load wins. Suppose they cannot move the load? When you finish this section, you should be able to:

☐ **A.** Explain what happens when a *force* is applied to an object.

☐ **B.** Identify examples of work being done.

☐ **C.** Describe how *force* is measured.

A **force** is a push or a pull on something. Scientists say that a *force* is only part of what is needed to do work. Work is done only when a force moves something. If you push on a wall, you are not doing any work because the wall is not moving. But if you push a chair and the chair moves, you are doing work.

The girl in picture 1 is not doing any work. She is holding a large rock. Her arms may get tired. But the rock is not moving. Work is done only when an object is moved. Look at the picture of the baseball player. He is about to throw the baseball. Is he doing work? As in picture 1, the answer depends on both force and motion. Now let's look at another example.

Force: A push or a pull on something.

3

In the third picture, the boy is reading a book. He probably is studying hard, but a scientist would say that the boy is not doing any work. Why would the scientist say this?

Scientists can measure the amount of work being done. To do this, they must know how much force is being used. A *spring scale*, like the one shown here, is used to measure force. If you pull on the scale, the spring will stretch. The harder you pull, the more the spring will stretch.

A spring scale can also be used to measure weight. When we measure weight, we are really measuring a force. Each object that we weigh pulls with a certain force on the spring. A heavy object pulls with more force than a light object. So a heavy object will stretch the spring more than a light object. A 1-kilogram (2-pound) book would stretch the spring twice as far as a 0.5-kilogram (1-pound) book.

ACTIVITY

How can force be measured?

A. Gather these materials: 30-cm (12 in.) ruler, 15-cm (6 in.) string, medium-sized rubber band, 4 paper clips, index card, scissors, 5 books, and box of chalk.

B. Make a scale to measure force using the ruler, paper clips, and a rubber band. Tape the index card onto the ruler. Mark off the card in centimeters.

C. Pile 5 books on your desk. Tie string around a box of chalk. Attach the box of chalk to the force scale. Lift the box straight up to the top of the book pile.

1. What was the force needed to lift the box?
2. What was the distance the box moved off the table?

Section Review

Main Ideas: Work is done when a force is used to move an object. Scientists define a force as a push or pull on something. A spring scale can be used to measure force.

Questions: Answer in complete sentences.

1. What is meant by a force?
2. What two things are needed to do work?
3. How do scientists measure force?
4. Find pictures in this chapter that show people doing work.

Have you ever driven on a steep mountain road? When a road is built over a mountain, it is not straight. As shown in the picture, the road curves from side to side. This curving makes the road longer. But it also makes it easier for cars to get over the mountain. Why? When you finish this section, you should be able to:

☐ **A.** Explain how a *machine* makes work easier.

☐ **B.** Explain how an *inclined plane* is used to do work.

Each day we do work of many kinds. We carry out the trash. We turn on the faucet in the kitchen sink. Can you name some other examples of work? The pictures that follow show work being done. What do you think makes the work easier in each case?

Anything that people use to make work easier
is called a **machine**. Perhaps when you hear the
word *machine*, you think of a crane or a car engine. Do you also know that many of the things
you use every day are simple machines? The
bottle opener in the picture above is a simple
machine. When you use a tilted board to move a
box, you are using a simple machine. How many
kinds of simple machines are there? How do
machines make work easier? Let's find out!

The person loading the car is using a simple
machine called an **inclined plane** (inn-**klind**
plane). An *inclined plane* is a slanted surface. It
is often called a ramp, a slope, or a hill.

Inclined planes make doing work easier. The
inclined plane shown above makes the work of
lifting the barrel easier. You have to push the
barrel farther, but you don't need as much force.

**Machine: Anything
that makes work
easier.**

**Inclined plane: A
slanted surface, which
is sometimes called a
ramp.**

If you have to use a lot of force, you get tired. A machine, like the inclined plane, helps you use less force to do work. Even though it might take a little longer to do the work, you won't be as tired. Pushing an object up a ramp is easier than lifting it straight up. Inclined planes are all around you. Ramps are used to make it easier for people in wheelchairs to enter buildings. The mountain road shown on page 178 is also a type of inclined plane.

The angle of an inclined plane affects the force needed to move an object. Look at the two ramps shown below. A spring scale is shown attached to each brick. Notice that less force is being used to move the brick on the left. The angle or slant of this inclined plane is less. Since the inclined plane is not as steep, it takes less force to move the brick.

ACTIVITY

How does an inclined plane help you move an object?

A. Gather these materials: a spring scale, 1 board, toy car or truck, and 3 books.

B. Copy the chart shown. Use it to record your results.

C. Make an inclined plane by placing the board on 1 book.

D. Use the spring scale to pull the toy car up the inclined plane. Record the force needed to move the car.

E. Change the angle of the inclined plane by adding the second book. Repeat step D.

F. Make the angle of the inclined plane steeper by adding the third book. Repeat step D.

    1. How does the angle of an inclined plane affect the force needed to move an object?

Number of Books	Force
1	
2	
3	

When work is done, the object is moved through a distance. Look at the truck on page 182. There are two ways for the driver to get the barrel into the truck. The driver can lift the barrel straight up. In this case, the distance is 1

181

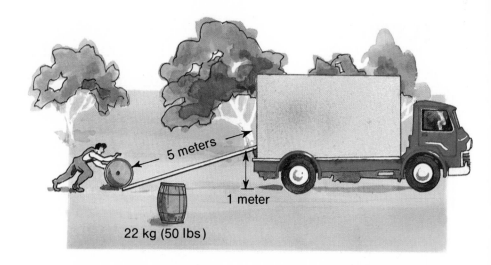

5 meters

1 meter

22 kg (50 lbs)

meter (3 feet). But to do this, he must use a lot of force. The other way to get the barrel into the truck is to use the ramp. The ramp is 5 meters long (about 15 feet). The driver moves the barrel a longer distance but uses less force.

Section Review

Main Ideas: Machines make work easier. They do not change the amount of work done. Inclined planes are simple machines that make work easier.

Questions: Answer in complete sentences.

1. How does a machine affect the work done?
2. What difference would it make if you walked up a flight of stairs instead of a ramp?
3. How could an inclined plane make moving a chair into a truck easier?
4. Suppose you have to walk up a very steep hill. What path would you take to make it easier going up the hill?

Hi
Person

Look at the picture of the winding staircase. There are over a hundred steps in this staircase. If you had a ladder the same height as the winding staircase, which would you rather climb: the staircase or the ladder? Why? When you finish this section, you should be able to:

☐ **A.** Explain why a *screw* and a *wedge* are kinds of inclined planes.

☐ **B.** Identify objects that are kinds of *screws* and *wedges*.

Many of the things in your home are held together with inclined planes. Also, many of the tools that we use to make and build things are hidden inclined planes.

The stairway in the picture on page 183 is a simple machine. It is a curved inclined plane. Do you recall how inclined planes make work easier? Instead of lifting an object up, the inclined plane allows you to move it along a slope. It is much easier to walk up a curved stairway than to climb straight up a ladder. You walk farther, but you use much less force.

Screw: An inclined plane that winds around in a spiral.

A winding stairway is a type of simple machine. It is called a **screw**. A *screw* is an inclined plane that winds around in a spiral. As a screw is pushed and turned into a board, its inclined plane moves through the wood.

Imagine being small enough to walk up a screw's winding edge. Wouldn't that be like walking up a winding stairway or a winding mountain road? What other types of screws can you think of? You cannot lift a car using only your arms. But a jackscrew can lift a car easily. As you turn the handle, the screw lifts the car.

Another kind of inclined plane that is a simple machine is the **wedge** (**wej**). A *wedge* is two inclined planes joined to form a sharp edge. Wedges are used to cut or break things apart. Knives and axes are wedges.

Look at the picture below. Imagine hitting the flat side of the wedge. Follow the path of the arrows. You see that your force becomes focused at the sharp edge of the wedge. Wedges focus forces to their sharp edges the way lenses focus light to one point. This makes it easier to force the wood apart. When you hit the wedge with a hammer, the hammer acts as a lever.

A nail is a pointed wedge that forces wood apart. Forks and needles are also wedges. When a wedge, such as an ax, is used, the wood moves along the wedge. A thinner ax will go into wood easier. It is like a shallow inclined plane. You do not use as much force to move objects along it.

Wedge: Two inclined planes joined together to form a sharp edge.

wedge ——

ACTIVITY

How is a screw like an inclined plane?

A. Gather these materials: 2 screws, 2 sheets of paper cut into triangles, crayon, 2 pencils, tape, screwdriver, and a wood block.

B. Look at the two screws.
   1. Which one will be easier to screw into the wood?

C. With the screwdriver, turn each screw into the wood.
   2. Which one was easier to screw in?

D. With the crayon, make a line along the slanted edge of each triangle. Note that both look like inclined planes.

E. Wrap each triangle around a pencil. Fasten each with tape. Count the number of turns on each pencil.
   3. Which one has more turns or threads?
   4. Which inclined plane had a smaller angle?
   5. How is a screw like an inclined plane?

Section Review

Main Ideas: The screw and the wedge are both simple machines. Both are inclined planes.

Questions: Answer in complete sentences.

1. How is a screw like an inclined plane?
2. Identify each of the following as either a screw or a wedge: (a) nail; (b) fork; (c) ax; (d) needle; and (e) winding staircase.
3. Which of the wedges shown in the diagram would be easier to drive into a block of wood? Explain.

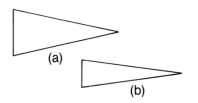

People in Science

on the lever bike move up and down. This means that the rider's legs move less. The rider uses less force to pedal. The force of the rider's feet on the pedals is carried to the rear wheel by levers. The rider can change speeds on hills by changing the length of the levers. Most bicycles have 10 speeds. The lever bike has 29 speeds!

The earliest bicycle was built in about 1816 in Germany. In 1886, a modern bicycle was built in England. Bicycles have not changed much from that time up until the invention of the lever bike.

Seol Man Taik

Seol Man Taik is a Korean inventor. He has made a new kind of bicycle. It is called the lever bike. Bicycle pedals usually move in circles. But the pedals

11-4.

The Lever

The board in the picture is resting on a small log. The smaller person wants to lift the larger person off the ground. She can stand either on the spot marked *x* or the spot marked *y*. Where do you think she should stand? Why? When you finish this section, you should be able to:

☐ **A.** Identify objects that are *levers*.

☐ **B.** Explain what a *lever* is and how it can make work easier.

☐ **C.** Show where to put the *fulcrum* when you want to move a heavy load.

Can you guess where the girl in the picture should stand? She should stand on the spot marked *y*. How will this make it easy for her to lift the other person off the ground?

A simple machine with a bar resting on a turning point is called a **lever** (**lee**-ver). The bar may be a stick, a rod, or a board. A seesaw is a *lever*. The turning point of a lever is called the **fulcrum** (**ful**-crum). Can you find the *fulcrum* on the seesaw? Let's find out how levers make work easier.

The girls in the picture above are using a lever to move a heavy rock. The tree branch is the bar. The spot where the branch rests against the box is the fulcrum. The girls push down on one end of the branch. The other end moves up and lifts the rock. Suppose the girls tried to move the rock without the lever. They would need a great deal of force to move the rock. They might not be able to move the rock at all. What happened when they used the lever? They pushed the lever down a great distance but used very little force. Like inclined planes, levers allow you to use less force to move things.

You can lift heavy loads with a lever. To do so, you need to know where to put the fulcrum. The diagram shows three levers. With which lever could you lift the heaviest load?

Lever: A bar resting on a turning point.

Fulcrum: The turning point of a lever.

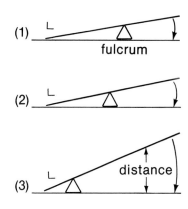

ACTIVITY

How is the force of a lever related to the location of the fulcrum?

A. Gather these materials: box of paper clips, flat-sided pencil, 2 small paper cups, small box, tape, and wooden ruler (30 cm).

B. Label 1 cup L for load. Label the other cup F for force.

C. Tape a cup to each end of the ruler. Place a pencil on top of a small box.

D. Put 10 paper clips into cup L. Place cup L 8 cm from the fulcrum. Put enough paper clips into cup F to balance the load. Record your results in a chart like the one shown here.

    1. How many paper clips did it take to balance the load?

E. Remove the paper clips from cup F. Repeat step D with the fulcrum at 12, at 15, and at 18 cm.

    2. Did you use more or less paper clips as the fulcrum got farther from the load?

    3. How does the location of the fulcrum affect the load a lever can lift?

Fulcrum at	Force*
8 cm	
12 cm	
15 cm	
18 cm	
*(Number of paper clips)	

190

A hammer can also be used as a lever. You can use a hammer to pull a nail out of a wall. The fulcrum is at the head of the hammer. The whole handle length is the lever. You can grasp the hammer at any spot along the handle. Where would you grasp it to make pulling the nail out easiest?

Remember that the closer the fulcrum is to the load, the less force you have to use.

Section Review

Main Ideas: Levers decrease the force needed to move things. The closer the fulcrum is to the load, the less force you need to move the load.

Questions: Answer in complete sentences.

1. Which of the levers shown will make lifting a load easier?
2. Make a diagram showing the fulcrum, lever, load, and force. Label each part.
3. The drawing shows two can openers. Which one would make it easier to open a can?
4. Where is the fulcrum on the can opener?

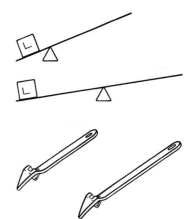

CHAPTER REVIEW

Science Words: Match the terms in column A with the definitions in column B.

Column A	Column B
1. Screw	a. The turning point of a lever
2. Lever	b. Two inclined planes
3. Force	c. A winding inclined plane
4. Fulcrum	d. A push or a pull
5. Machine	e. Makes work easier
6. Wedge	f. A bar resting on a point
7. Inclined plane	g. A slanted surface

Questions: Answer in complete sentences.

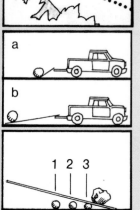

1. What is work?
2. In which of the following examples is work *not* being done: (a) pushing on a wall that does not move; (b) lifting a box off the floor; (c) watching television; and (d) throwing a basketball?
3. How can force be measured?
4. How does a machine help us do work?
5. The picture shows the path cows take over a steep hill. Why do the cows choose a longer, zig-zag path?
6. On which inclined plane will it be easier to push a barrel up to a truck? Explain.
7. How are screws and wedges alike?
8. How is a screw like a winding staircase?
9. At which spot would you put a fulcrum to make lifting the rock easiest? Why?
10. What are two examples of levers that you might use at home?

CHAPTER 12

MACHINES WITH WHEELS

Machines with wheels can help you do many jobs. One kind of machine is made of wheels and ropes. This machine helps people lift heavy loads. Do you know what this machine is? When you finish this section, you should be able to:

12-1.

Pulleys

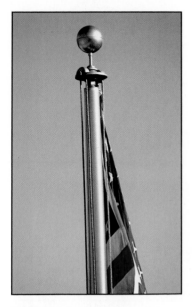

Pulley: A simple machine that is a wheel with a rope moving around it.

Fixed pulley: A pulley that stays in place as the load moves.

Movable pulley: A pulley that moves with the load.

☐ **A.** Explain what a *pulley* is and give some examples.

☐ **B.** Explain why a *fixed pulley* makes work easier.

☐ **C.** Explain why a *movable pulley* makes work easier.

A **pulley** is a simple machine that is a wheel with a rope moving around it. The rounded edge of the wheel usually has a groove in it. The groove keeps the rope from slipping off the *pulley*. The picture above shows a pulley being used to raise a flag. The person raising the flag is pulling down on the rope. The rope moves around the wheel and pulls the flag up. This pulley stays in place as the load moves because the wheel is fastened to one spot. A pulley that does not move is called a **fixed pulley**.

A *fixed pulley* changes the direction of the force used. You pull down and the load is pulled up. The amount of force used to lift the load is the same needed to lift it straight up. The work is easier because, as with the lever, pushing or pulling down is often easier than pulling something up.

The pulley shown in the margin is not fastened to one spot. It is attached to the load. It moves with the load. As the load is lifted, the pulley is also lifted. A pulley that moves is called a **movable pulley**. When you use a *movable pulley*, you pull up to lift the load. But your work is easier. You use less force to do the lifting.

Often, a fixed pulley and a movable pulley are used together. This makes the force needed to lift a load even less. The ropes between the pulleys share the force. You can pull on one rope with less force, but for a longer distance. The crane in the picture below uses fixed and movable pulleys.

Eight ropes are moving between the pulleys. They share the load among them. Each rope lifts only one eighth of the load. The crane must then pull 8 meters (26 feet) of rope to lift the load 1 meter (3 feet). Suppose the crane lifts the load 15 meters (50 feet). How much rope would it have to pull?

ACTIVITY

How do pulleys make work easier?

A. Gather these materials: small pulley, meter stick, string, pail, sand, and spring scale.

B. Fill the pail about 1/4 full of sand. Lift the pail with the spring scale.

    1. What force was needed to lift the pail?

C. Attach 1 end of the string to the meter stick, as shown. Run the string through the pulley. Attach the free end of the string to the spring scale. Hook the pail onto the pulley and lift the pail.

    2. What was the force needed to lift the pail using the pulley?

    3. How does this force compare to the force needed to lift the pail without the pulley?

    4. How does a pulley make work easier?

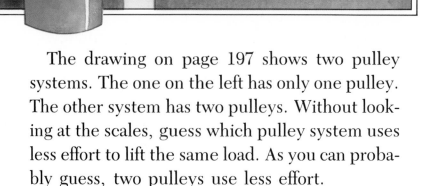

The drawing on page 197 shows two pulley systems. The one on the left has only one pulley. The other system has two pulleys. Without looking at the scales, guess which pulley system uses less effort to lift the same load. As you can probably guess, two pulleys use less effort.

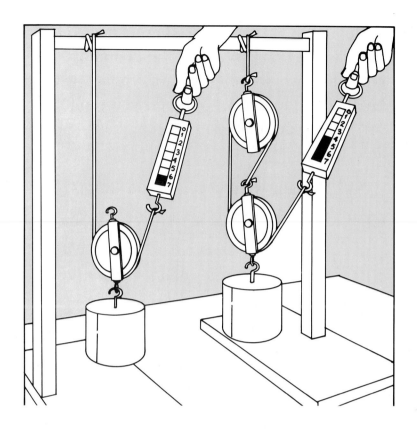

Section Review

Main Ideas: Fixed pulleys make work easier by changing the direction of force. Movable pulleys make work easier by making the force less. Often, fixed pulleys and movable pulleys are used together.

Questions: Answer in complete sentences.

1. Draw a picture of a pulley. Explain how it can be used to lift a heavy load.
2. What is a fixed pulley and a movable pulley?
3. Draw a picture to show how a fixed pulley and a movable pulley could be used together.
4. Which of the pulley systems shown at the top of this page makes work easier? Why?

12-2.

The Wheel and Axle

A doorknob is part of a simple machine that unlocks the door. Without the doorknob, opening a door would be very hard. In fact, it would be almost impossible. Why do you think this is so? When you finish this section, you should be able to:

☐ **A.** Explain what a *wheel and axle* is.

☐ **B.** Describe how a *wheel and axle* makes work easier.

☐ **C.** Identify examples of a *wheel and axle*.

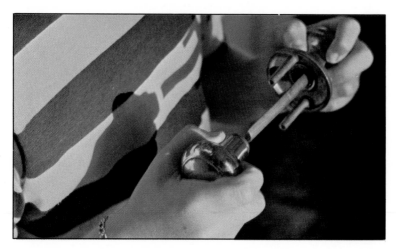

Look at the picture above. Have you ever taken a doorknob apart? The round knob is attached to a rod. It is very hard to turn the rod without turning the knob. Why do you think this is so?

A wheel that turns on a rod is called a **wheel and axle** (ak-sell). A doorknob and the rod it is attached to is a *wheel and axle*. The doorknob is the wheel. The rod is the axle. An axle usually goes through the center of a wheel. When the wheel turns, the axle turns. When the axle turns, the wheel turns.

When you turn a doorknob, you are using force to turn the wheel part of a wheel and axle. You move your hand more than you would if you just turned the rod. But you turn with a lot less force. Your work is made easier.

The wheel of a wheel and axle may be hard to recognize. You can't always see the wheel. The handle of a food grinder is a type of wheel and axle that does not seem to have a wheel. The wheel is the circle made in the air when you turn the handle.

Wheel and axle: A simple machine that is a wheel that turns on a rod.

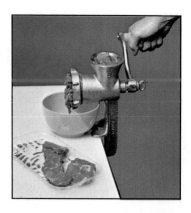

199

ACTIVITY

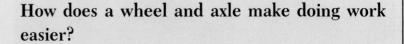

How does a wheel and axle make doing work easier?

A. Gather these materials: 2 screwdrivers with different-size handles, 2 screws (same size), wood block, and metric ruler.

B. Measure the thickness of the handle of each screwdriver.

C. Use the screwdriver with the thinner handle to screw one of the screws into the wood.
   1. How many turns did you make?

D. Use the screwdriver with the thicker handle to screw in the second screw.
   2. How many turns did you make?
   3. With which screwdriver was it easier to turn the screw?
   4. Why was there a difference between the 2 screwdrivers?

There are many other examples of the wheel and axle. Look at the two screwdrivers shown here. They are both the same length. But the handles are different sizes. The handle of a screwdriver is the wheel. The metal part is the axle. Which screwdriver would be easier to use? Since it is easier to turn a larger wheel, the screwdriver with the larger handle is easier to use. You have to turn it more, but each turn is easier to make.

Some other examples of the wheel and axle are shown above. Look at each one. You should be able to find the wheel and axle shown in each of the pictures.

Section Review

Main Ideas: A wheel turning on a rod, or axle, is a kind of simple machine. The axle usually goes through the center of the wheel.

Questions: Answer in complete sentences.

1. What are two examples of a wheel and axle in your home?
2. Why is unlocking a door easier with a door-knob than without it?
3. Why do you think buses have very large steering wheels?
4. Look at the screwdriver shown in the margin. Which part can be thought of as a wheel? Which part is an axle?

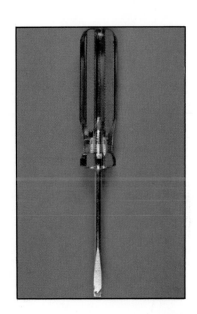

12-3.

Gear Wheels

These children are riding 10-speed bicycles. They can change gear by using a gear shift. If they are going up a steep hill, they change to a gear that will make it easier to pedal. What speed should they use to go up a steep hill? When you finish this section, you should be able to:

☐ **A.** Identify machines that contain *gears*.

☐ **B.** Explain how *gears* make work easier.

☐ **C.** Explain why chains are used to connect *gear wheels*.

When you ride a bicycle, you are doing work. The pedals are attached to a large wheel that has

points, or teeth. The back wheel is attached to another toothed wheel. A wheel with teeth is called a **gear** or **gear wheel**. *Gear wheels* are usually attached to other gear wheels. On bicycles, the gears are joined by a chain. The teeth fit into open places in the chain. When you pedal, you turn the larger gear wheel. This turns the chain, which turns the smaller gear wheel. Your bicycle goes forward.

Gear wheel: A simple machine that is a wheel with teeth.

The boy in the picture is using a machine called an egg beater to mix frosting. The turning handle is attached to a larger gear wheel. The mixing blades are attached to a smaller gear wheel. The larger gear wheel turns the smaller gear wheel. In an egg beater, the gear wheels fit without a chain. How is work made easier when large gear wheels turn small gear wheels?

Gear wheels, like the one on an egg beater, make work easier. When you turn the large wheel around, the smaller wheel makes more turns. The blades also turn faster than the larger wheel.

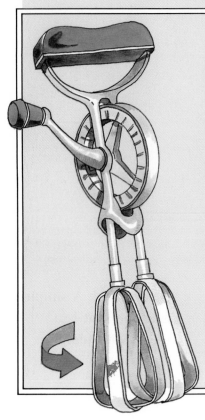

How does a gear wheel work?

A. Gather these materials: crayon and egg beater.

B. Put a crayon mark on 1 blade of an egg beater.

C. Slowly turn the large gear wheel 1 complete turn. Use the mark on the blade to count how many turns the blade made.

D. Repeat step C, turning the gear wheel 2 times, and then 3 times.

    1. How many times did the blade turn each time?

    2. Did the blade turn faster or slower than the large gear wheel?

Do you know why the gear wheels on a bike are joined by a chain? Look at the drawings below. The gear wheels without the chain turn in opposite directions. Wheels with the chain turn in the same direction. If the gear wheels on a bicycle were joined without a chain, you would have to pedal backward to go forward.

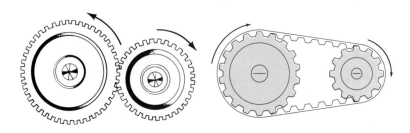

These pictures show a close-up of the gear and wheels on a bike and those inside a watch. The gear wheels on the rear wheel of the bike are of different sizes. The watch also has different-size gear wheels. Why do you think the watch has different-size wheels?

Section Review

Main Ideas: Large gear wheels that turn small gear wheels make work easier. They increase the speed of the small gears and the number of times they turn. Two gear wheels that are joined by a chain turn in the same direction.

Questions: Answer in complete sentences.

1. How do gear wheels make work easier?
2. How can two gear wheels be made to turn in the same direction?
3. In the picture of the bike, which size gear wheel would make pedaling easier?
4. What does it mean when we say a bicycle has 10 speeds?

12-4.

Compound Machines

Have you ever taken your bike apart to clean or repair it? If you have, then you know that it is not as simple as it looks. A bicycle is a machine. Most machines are made of two or more simple machines that work together. What are other examples of such machines in your home? When you finish this section, you should be able to:

☐ **A.** Identify and give examples of *compound machines*.

☐ **B.** Identify the simple machines that are in a *compound machine*.

A machine made up of two or more simple machines is called a **compound machine** (kom-pound). A bicycle is a *compound machine*. The word *compound* means to mix or put together. A bicycle is made of many simple machines. The front and back wheels are wheel and axles. The handlebar and pedals are also wheel and axles. The screws that hold the seat and handlebar in place are hidden inclined planes. The toothed wheels joined by the chain are gear wheels.

The tool shown above is a compound machine called a hand drill. A hand drill is made of three simple machines. Do you know what they are? The handle of the drill is a wheel turning on an axle. When you turn the handle, you are turning a rod that moves in a circle. The tip of the drill is spiraled like a screw. It is an inclined plane. The sides of the tip are wedge-shaped. They are sharp and cut like knives.

Compound machine: A machine made of two or more simple machines.

207

ACTIVITY

What are the parts of compound machines?

A. Gather these materials: scissors, hand drill, toy crane, and pencil sharpener.

B. Copy the chart shown below. Use it to record your observations.

Compound Machine	Inclined Plane	Screw	Wedge	Lever	Pulley	Wheel and Axle
Scissors						
Hand drill						
Toy crane						
Pencil sharpener						

C. Look at each compound machine. Find as many simple machines in each as you can. Describe each in your chart.

1. What are compound machines made of?
2. How many simple machines did you find in each compound machine?

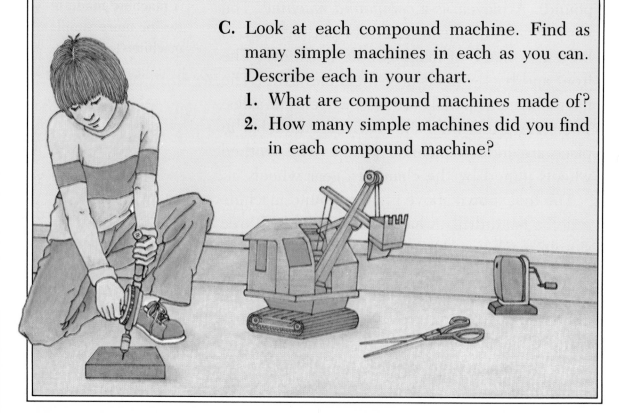

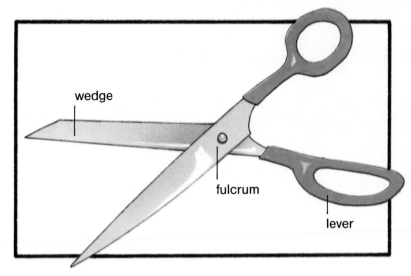

Scissors are also compound machines. A pair of scissors is made of two levers. The fulcrum is the place where the levers are joined. Each lever also has a wedge-shaped cutting edge.

Another example of a compound machine is a movie projector. How many simple machines can you find in the movie projector?

Section Review

Main Ideas: Compound machines are machines made of two or more simple machines. Bicycles, hand drills, and scissors are examples of compound machines.

Questions: Answer in complete sentences.

1. What simple machines can you find in the machine shown in the margin?
2. What are three examples of compound machines?
3. What is the advantage of a compound machine over a simple machine?

CHAPTER REVIEW

Science Words: Copy the numbered letters and spaces on paper. Use the clues to help you identify the science terms.

1. __ u __ __ __ __

2. __ __ x __ __ __ __ __ __ __ y

3. __ __ v __ __ __ __ __ __ l __ __ __

4. w __ __ __ __ __ __ __ __ __ __ e

5. __ e __ __ __ __ e __ __

6. __ __ __ p __ __ __ __
   __ __ __ __ i __ __

Clues

1. A wheel with a rope moving around it
2. A machine that stays in place as the load moves (two words)
3. A machine that moves with the load (two words)
4. A wheel that turns on a rod (three words)
5. A wheel with teeth (two words)
6. A machine made of two or more simple machines (two words)

Questions: Answer in complete sentences.

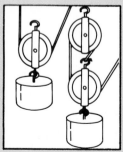

1. Draw a picture showing the difference between a fixed and a movable pulley.
2. Which of the pulleys shown here would make work easier to do?
3. How does a wheel and axle make work easier?
4. Draw a picture showing how the gear wheels on a bike are joined with a chain.

210

CHAPTER 13

FRICTION AND WORK

In January 1973, a freezing storm struck Atlanta, Georgia. This storm was later called the "Great Atlanta Ice Storm." Cars, trucks, and buses slid all over the icy streets. When you finish this section, you should be able to:

13-1.

What Is Friction?

A. Explain how *friction* affects the force needed to move objects.

B. Identify helpful and harmful effects of *friction*.

These children are going to play a game of tug-of-war. Do you think it will be a fair game? Do you think one group has an advantage?

If you said it would be unfair, you were correct. The children on the ice will be pulled easily. They will not be able to move the children standing on the ground. The reason for this is **friction** (**frik**-shun). *Friction* is a force that acts when two objects rub or slide over one another. Friction keeps objects from moving. The friction between the ground and the children's shoes keeps them from slipping.

The friction is less between the shoes and the ice. Thus, the children on the ground will win easily.

Friction: A force caused when one object rubs against another object.

Smooth surfaces make less friction than rough surfaces. It takes more force to move an object over a rough surface than a smooth surface.

For example, suppose you want to push a heavy box across the ground. There is a lot of friction between the box and ground. This makes pushing the box very hard. Putting a blanket under the box reduces the friction. The blanket slides easily over the ground. Now moving the box is easier. The box would also slide more easily over a polished floor. Why?

Friction can be very helpful. One of the first uses of friction was to make a fire. Early humans found that rubbing two sticks together caused the sticks to become hot. The friction between the sticks could make them hot enough to cause a fire. In bicycle brakes, the rubber of the brakes rubs against the metal of the wheel. The friction between these two surfaces causes the wheels to slow. Can you think of some other helpful uses of friction?

ACTIVITY

How is friction affected by different surfaces?

A. Gather these materials: shoe box, sand, sandpaper, wax paper, and spring scale.

B. Fill the box 1/2 full of sand. Attach the spring scale to the box.

C. Pull the box across a piece of wax paper.
   1. What force is needed to move the box?

D. Pull the box across a piece of sandpaper.
   2. What force is needed to move the box?
   3. Over which surface was it harder to pull the box?
   4. What effect does the type of surface have on friction?

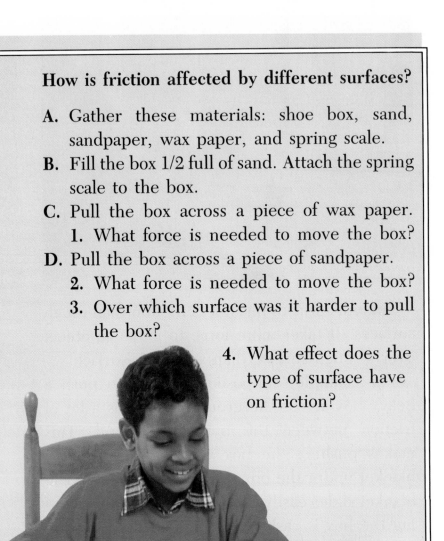

Friction can cause problems, too. Too much friction makes heavy objects hard to move. When objects rub against each other, the surfaces are worn away. Look at your shoes. Friction with the ground causes the heels and soles to wear away. Friction also wears away the parts of machines. Belts connect the moving parts inside a car engine. They are worn away by friction. What are some other problems caused by friction?

Too little friction is a big problem on icy roads. To solve this problem, we must increase the friction between the ice and the tires of cars. There are many ways to do this. Chains can be put on the tires. The chains make the surfaces of the tires rougher. Friction is increased. You can also throw sand on the ice. The sand grains make the icy surface rougher. Again, the friction is increased. Can you think of other times when you would want to increase friction?

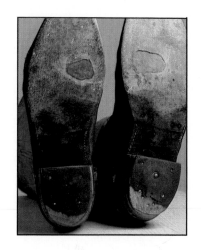

Section Review

Main Ideas: Friction is present when two surfaces rub or slide over one another. Brake systems are based on the force of friction. Too little friction can make it hard for us to move around.

Questions: Answer in complete sentences.

1. A girl rolls a ball across a field. The ball slows down and stops. Why does it do this?
2. Why do your hands get warm when you rub them together?
3. Why do we cut grooves in the cement on highways?
4. List three ways in which friction is helpful to us.

13-2.

Friction and Wheels

The picture shows how some of the work of building the pyramids might have been done. Thousands of heavy stone blocks had to be moved into place. The builders probably put logs under the stones to reduce friction between the stones and the ground. When you finish this section, you should be able to:

☐ **A.** Explain how wheels and other round objects reduce friction.

☐ **B.** Explain how wheels make machines more *efficient*.

Can you picture a wheelbarrow with a square wheel? It would be very hard to use to move a load of dirt. The round wheel was one of the earliest human inventions. People found that wheels make it easier to move things. A wheel reduces friction because only a small part of it touches the ground at one time.

Wheels are not the only round objects that reduce friction. **Ball bearings** are also used to reduce friction in machines. A *ball bearing* is a smooth ball made of metal. Because the balls are smooth, they can roll next to each other easily. Ball bearings are used in roller skates and skateboards to make the wheels turn more smoothly. Look at the picture in the margin, which shows the inside of a wheel. The small metal balls are ball bearings. They reduce the friction. This allows the wheel to spin smoothly.

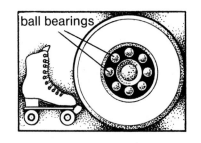

Ball bearings: Small metal balls used to reduce friction.

ACTIVITY

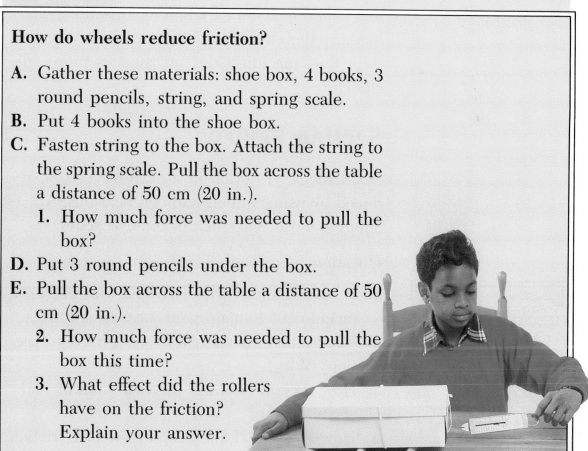

How do wheels reduce friction?

A. Gather these materials: shoe box, 4 books, 3 round pencils, string, and spring scale.

B. Put 4 books into the shoe box.

C. Fasten string to the box. Attach the string to the spring scale. Pull the box across the table a distance of 50 cm (20 in.).

   1. How much force was needed to pull the box?

D. Put 3 round pencils under the box.

E. Pull the box across the table a distance of 50 cm (20 in.).

   2. How much force was needed to pull the box this time?

   3. What effect did the rollers have on the friction? Explain your answer.

Efficiency: The amount of work done by a machine compared to the amount of work put into it.

Machines are more **efficient** (ee-**fish**-ent) if we can reduce friction. The *efficiency* of a machine depends upon how much work we get out of it. We never get out of a machine as much work as we put into it. Some of the work we put into the machine is lost to friction. Wheels and ball bearings can help reduce friction in a machine. They increase the efficiency of the machine. Which of these two machines do you think is more efficient?

Section Review

Main Ideas: Friction in a machine can be reduced by using wheels and ball bearings. A machine is more efficient if friction is reduced.

Questions: Answer in complete sentences.

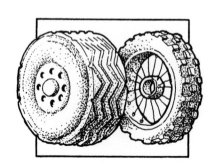

1. Which of these wheels would you use on a racing car? Explain your answer.
2. What effect does a ball bearing have on friction? Why?
3. How do ball bearings increase the efficiency of a machine?
4. How do you think people reduced friction before the invention of the wheel?

Reducing Friction

The boat in the picture is called a **Hovercraft** (**huv**-er-kraft). Unlike other boats, a *Hovercraft* does not move through water. It rides on a layer of air just above the water. A Hovercraft can move very fast. Why do you think this is so? When you finish this section, you should be able to:

☐ **A.** Explain how *lubrication* reduces friction.

☐ **B.** Explain the effect of lubricants on the efficiency of machines.

There is a great deal of friction between the bottom of a boat and the surface of the water. Speedboats with very large engines can move faster than some cars. But usually boats do not travel at high speeds. This is because friction is greater in the water than on land. The water touches the whole bottom surface of the boat. If

Hovercraft: A boat that rides on a cushion of air to reduce friction with the water.

219

you could reduce the amount of water touching the boat, it would go faster. A clever way to do this is to lift the boat. Lifting the boat just above the water reduces friction. The Hovercraft blows air down. This cushion of air lifts the boat up. There is a slight space between the boat and the water. Thus, the Hovercraft is able to travel quite fast.

The problem of friction cannot always be solved by separating the two surfaces. In most machines, the parts must touch each other. We have to find other ways to reduce friction. **Lubrication** (loo-bri-**kay**-shun) is the method used in many machines.

Lubrication reduces friction by putting a material like oil between the parts of a machine. For example, rub your hand against your desk. You should feel the friction between your hand and the desk. Now wet your hand with a little water. Your hand should slide easily across the desk. The water acts as a lubricant. A lubricant allows two surfaces to rub smoothly against one another. This reduces friction.

The efficiency of a machine is less if it has to do extra work against friction. But when lubrication helps make the friction less, the machine becomes more efficient. In the following two examples, lubrication helps the machines run more efficiently.

The girl in the picture is putting some oil on the chain of her bike. The oil will lubricate the moving parts of the bike. Friction will be reduced. The bike will ride smoother.

Lubrication: A way of reducing friction by using a liquid on the moving parts.

ACTIVITY

How do lubricants reduce friction?

A. Gather these materials: 4 marbles, liquid soap, and a paper towel.

B. Rub your hands together. Do this for about 10 seconds.

  1. What did you feel?

C. Put a small amount of liquid soap in your palm. Rub your hands together. Wipe your hands with a paper towel.

  2. What happened when you rubbed your hands together this time?

  3. Why was there a difference?

D. Hold the 4 marbles. Rub your hands together again.

  4. How does this compare with rubbing your hands together without a lubricant?

E. Pour a small amount of lubricant (liquid soap) on the marbles. Rub your hands together while holding the lubricated marbles.

  5. How did the lubricant affect the friction of the marbles?

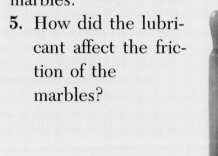

A car engine also needs oil. The engine contains pistons that move up and down. When the pistons are oiled, they move up and down smoothly. Checking the oil level in a car is very important. Without the oil, the pistons will stop. The engine will not run efficiently.

Many kinds of materials are used as lubricants. Look at the materials shown above. In what way could you use each as a lubricant?

Section Review

Main Ideas: Friction is caused when the moving parts in a machine rub against each other. A lubricant reduces friction. When the friction is reduced, the machine is more efficient.

Questions: Answer in complete sentences.

1. What effect does oil have on the moving parts of a toy wagon?
2. Why do people have the oil in their car engines checked?
3. How could you increase the friction between two moving parts?

CHAPTER REVIEW

Science Words: Think of a word for each blank. List the letters **a** through **e** on paper. Write the word next to each letter.

The force caused when one object rubs against another is ___**a**___. Small balls used to reduce friction are called ___**b**___. To make a machine more ___**c**___, we try to reduce the friction between its moving parts. Often, this is done by ___**d**___. A liquid used to reduce friction is called a ___**e**___.

Questions: Answer in complete sentences.

1. Why does a hockey puck slide across the ice easily?
2. Why can you use a nail file to change the shape of your nails?
3. Name one harmful example of friction.
4. Explain how friction acts in each of the following examples: (a) car brakes; and (b) an icy road.
5. When would it be helpful to increase friction?
6. Why does a flat tire increase friction with the road?
7. How do wheels help reduce friction in a machine?
8. Without pedaling faster, what is one way to make a bike go faster?
9. How does a lubricant reduce friction?
10. Why do you think highways and roads are slippery after it rains?

How do scientists measure work?

A. Gather these materials: spring scale, string, meter stick, 3 pencils, milk carton, sand, 50-cm (20-in.) board, and 2 books.

B. Copy the chart shown here. Use it to record your results.

C. Fill the milk carton about 1/2 full of sand. Close the top. Fasten a piece of string to the carton.

D. Make an inclined plane by stacking the 2 books. Lay a board on them to form a ramp. Work is equal to the force times the distance. The formula for work is $W = F \times d$

E. Fasten the milk carton to the spring scale. Slowly pull the carton up the ramp. Record the force used and the distance up the ramp.

  1. Using the formula, how much work was done moving the milk carton? Record.

F. Put more sand in the milk carton until the carton is about 3/4 full.

  2. Do you think you will have to do more or less work now to pull the carton up the ramp?

G. Repeat step E.

  3. How much work did you do? Record.

  4. Was the amount of work different in the two trials? Why?

Load	Force	Distance	Work
1/2 full			
3/4 full			

Machinist ▶

Machinists make the metal parts from which machines are made. They must be skilled in using machine tools to cut metal into exact sizes and shapes. All the parts must be made so that the finished machine will be as efficient as possible. People who want to be machinists must first study with a master machinist.

◀ Mechanic

What do you do when something goes wrong with your car or lawnmower or dishwasher? You may be good at do-it-yourself repairs. Or you may take it to a mechanic. Mechanics usually study machines at a technical school. They must know how machines work and how to fix them when they break down.

ANIMAL AND PLANT POPULATIONS

UNIT 5

CHAPTER 14

LIVING THINGS

14-1.

The Biosphere

The earth is alive, from the bottom of the ocean to the tops of mountains. Living things are found on the edge of glaciers. Tiny creatures live in boiling hot springs. When you finish this section, you should be able to:

☐ **A.** Describe the parts that make up the *biosphere*.

☐ **B.** Explain the effect of the atmosphere and crust on the *biosphere*.

You can find forms of life in most places on the earth's surface. Forms of life are found (1) in soil, (2) in the ocean, (3) inside human beings, (4) on the land, (5) on plants, and (6) in the air.

Can you match the right picture with one of these forms of life?

Biosphere: The layer of earth that is made up of living things.

System: A group of things that act together.

Protists: Organisms that are neither plants nor animals.

All of the living things shown are part of the earth. This part of the earth is called the **biosphere** (by-ohs-sphere). Look at the picture below. The *biosphere* is in the middle of a "sandwich." The bottom of the sandwich is the outer layer of the earth. This layer is called the crust. The crust contains rocks and soil. The top of the sandwich is the atmosphere. Living things need the gases in the atmosphere to continue living. Animals use oxygen. Most green plants use carbon dioxide. Animals and plants also need the sun. The sun is the source of energy for the biosphere. Plants need sunlight to make food. Animals eat plants and other animals.

We might think of the biosphere as a **system**. The parts of a *system* work together. Look at the picture in the margin. It shows that the biosphere needs three things. They are sunlight, air, and the minerals in rocks and soil. The main forms of life are plants, animals, and **protists**. *Protists* are neither plants nor animals.

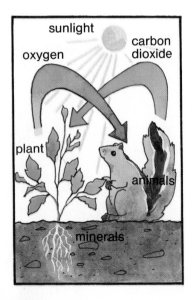

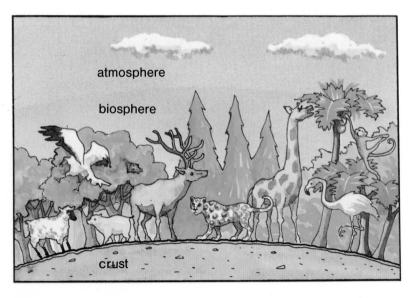

ACTIVITY

What makes up the biosphere?

A. Gather these materials: plastic shoe box or 1-l (1.1 qt) jar, pebbles, soil, grass seeds, small plant, and water.

B. Place the pebbles on the bottom of the box. Add a layer of soil about 4 cm (1.6 in.) thick.

C. Add the plant. Sprinkle the grass seeds in the soil. Add a small amount of water. Slightly wet the soil.

   1. What are the living things in the model?
   2. What part does the soil play in the model?
   3. What is the role of the air in the model? the role of the water?

Section Review

Main Ideas: The biosphere is made of animals, plants, and protists. Living things need the upper crust and the atmosphere.

Questions: Answer in complete sentences.

1. What is the biosphere?
2. How do the air and crust affect the biosphere?
3. What would happen to the biosphere if there were no sunlight?
4. Are the oceans part of the biosphere? Explain your answer.

231

14-2.

Is It Alive?

Pretend you are from another planet. You have a mission. It is to find out if there is life on the earth. How will you decide if something is alive? How are living things different from other things on earth? When you finish this section, you should be able to:

☐ **A.** Identify the four features of living things.

☐ **B.** Give examples of how *organisms* show these features.

The pictures you see below show two living things. Living things are called **organisms (or-gan-i-zums)**. The green *organism* is made of single cells. It is called **algae** (**al**-jee). *Algae* are protists. They float on ponds. The mushrooms in the second picture are a **fungus** (**fun**-giss). A *fungus* is a type of protist. The mold on rotting fruit is also a fungus.

Organism: A single living thing.

Algae: Very small protists made up of one or more single cells.

Fungus: A protist, such as a mushroom or a mold, that has no leaves or stems.

Some features are common to all living things. This is true for protists as well as for plants and animals. Four of these features are listed below and on page 234.

1. Living things **reproduce** (ree-pro-**doose**). All organisms come from organisms of the same kind. If living things could not *reproduce*, there would be no living things left on earth. Most single-celled organisms reproduce by splitting in half. Most plants reproduce by means of seeds.

2. Living things grow and change. When the foal shown here was born, it was less than half the size of its mother. When it becomes an adult, it will be the same size as its mother. Small trees will grow to be as tall as many old trees in the forest. Some living things change form as they get older. When the caterpillar becomes an adult, it will be a butterfly.

Reproduce: To make more of the same kind of organism.

3. Living things need food and water. Most animals spend much of their time looking for food. Some animals eat only plants. Others eat both plants and animals. Which of the animals shown here eats only plants? Which eats both plants and animals?

4. Living things need energy. An animal needs energy to run, swim, or fly. It gets the energy from the food it eats. Living things change food to energy inside their cells.

ACTIVITY

How can you tell if it is alive?

A. Gather these materials: piece of chalk, a rock, a cut flower, pencil, moss, and an ant.

B. Use the chart in Main Ideas below to decide if each object is a living thing.
  1. Which objects are living things?
  2. How can you tell a living thing from an object that is not living?

Section Review

Main Ideas: Features of living things:

1. They reproduce.	Cells split in half to make more cells.
2. They grow and change.	Animals and plants grow in size. Caterpillars change to butterflies.
3. They need food and water.	Animals eat plants.
4. They need energy.	Body cells change food to energy.

Questions: Answer in complete sentences.

1. Is a crystal a living thing? Explain your answer.
2. A student blows up a balloon and says, "It grew bigger. It is alive." Is the student right?
3. How can you tell that something is not alive?
4. Is a seed alive? Explain.

14-3.

Counting Living Things

Biologist: A scientist who studies the world of living things.

Population: A number of the same kind of organisms that live in the same place.

Suppose a disease is killing the trees in the picture. To keep track of the disease, you must count the trees. How will you count all the trees shown? When you finish this section, you should be able to:

☐ **A.** Describe a way to count a *population*.

☐ **B.** Explain why it is important to know the size of *populations*.

Biologists (bye-**oll**-oh-jists) study **populations** (pa-pew-**lay**-shuns) of living things. A *population*

is a number of the same kind of organism living in the same place. The picture in the margin shows a population of corn plants. What are some other populations?

Biologists can tell the size of a population by counting organisms. This helps them answer questions such as these:

1. Is the size of the population getting larger or smaller?

2. If the population is getting too small, should we try to protect the plants or animals?

3. If the population is getting too large, should we help make it smaller? Will there be enough food for all the living things?

Many people need to know the size of a population. Suppose you want to open a lumber company. You would want to know the number of trees in the area. A grocer would want to know the number of people in town. A teacher has to know how many students are in class. Can you think of any other examples?

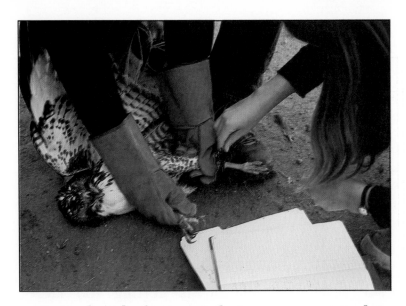

Animal and plant populations are counted in different ways. It may be hard to count animals. They move and are hard to find. They may be counted twice. They may live in places that are hard to get to.

Biologists have ways to mark the animals they count. Bands are put on the legs of some animals, such as birds. Other animals, such as rabbits, are marked on their ears. Radios are often put on animals like bears or deer. Biologists can follow the radio signals.

Plants do not move from place to place. So they are easier to count. Scientists do not count each plant in an area. They count the number of plants in a small plot of ground.

Suppose you wanted to count all the yellow flowers in a field. You would not have to count each flower. You could divide the field into squares. By counting the flowers in a square, you could get an idea of how many flowers are in the whole field.

7 yellow flowers x 16 squares =
7 x 16 = 112 flowers in the field

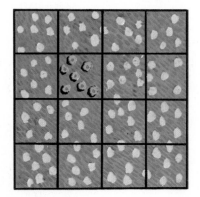

ACTIVITY

How is a population counted?

A. Gather these materials: 42-cm length of wire and metric ruler.

B. Copy the chart shown here. Use it to record your data.

C. Go outside to the roped-off area. Two bags of beans have been spread out in this area.

D. Bend the piece of wire to make a square 10 cm (4 in.) on each side. You will count the number of beans in this square.

E. Close your eyes. Drop the square on the ground. Count the number of beans in the square.

F. Repeat step E 4 times in other places.

G. Add the numbers in the count column to get the total count. Divide the total by 5 to find the average count.

   1. How could you find the total number of beans in the roped-off area?

Sample	Bean Count
1	
2	
3	
4	
5	
Total	
Average	

Section Review

Main Ideas: Biologists use many methods to count populations of plants and animals.

Questions: Answer in complete sentences.

1. How can you find the size of a population without counting each organism?
2. Why is it important to know the size of a population of whales?
3. How could you find out if a population was getting bigger or smaller?
4. Why is it harder to count animals than plants?

People in Science

Kes Hillman

Dr. Hillman is a biologist in Kenya, a country in East Africa. She studies the population of elephants in that country. Dr. Hillman's father was a pilot. Because of this, she became interested in flying. When she went to Kenya, Dr. Hillman found that flying helped her to study elephants. In an airplane, she could cover a large area. She could take pictures and count elephants more easily from the air. She could spot sick or hurt elephants. Her work helps keep the elephant population strong and healthy.

Dr. Hillman can also tell if elephants are being killed illegally. Each year, poachers kill many elephants for their tusks alone, from which luxury items are made.

CHAPTER REVIEW

Science Words: Match the terms in column A with the definitions in column B.

Column A	Column B
1. Biosphere	a. A group of the same kind of organism
2. System	
3. Protists	b. The living layer of earth
4. Organism	c. A scientist who studies living things
5. Algae	d. Things that work together to form a whole
6. Fungus	
7. Reproduce	e. To make more of the same kind of organism
8. Biologist	
9. Population	f. Single-celled plants
	g. A plant-like protist that has no leaves or stems
	h. Single-celled organisms
	i. A living thing

Questions: Answer in complete sentences.

1. In a drawing, show the biosphere, the atmosphere, and the crust.
2. How is the biosphere different from the other layers?
3. What might happen to the biosphere if the earth lost its atmosphere?
4. What are four features of all living things?
5. Why do living things need food?
6. Give one way in which living things reproduce.
7. Which would be easier: (a) counting birds in your yard, or (b) counting blue flowers in your yard?

CHAPTER 15

THE CYCLES OF POPULATIONS

15-1.

Animal Life Cycles

The giant sea turtle is ready to lay her eggs. She comes out of the sea and onto the beach. She digs a hole in the sand and lays her eggs. What happens to the baby turtles? When you finish this section, you should be able to:

☐ **A.** Describe and give examples of animal *life cycles*.

☐ **B.** Explain how birth, growth and development, and death affect *life cycles* of living things.

Below is an enlarged picture of very small animals. They are about the size of a comma. These animals live in salt water. They are called **brine shrimp**. The dots next to them are *brine shrimp* eggs. The eggs are very tiny. Brine shrimp reproduce by laying eggs. When the eggs hatch, the population changes size. New shrimp are added. The birth of animals helps populations grow and survive. What would happen to the brine shrimp population if the shrimp stopped laying eggs?

Brine shrimp: A very small animal that lives in salt water.

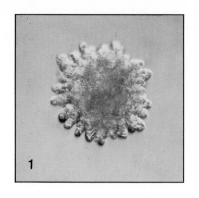

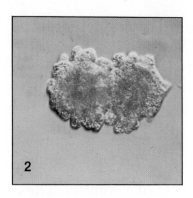

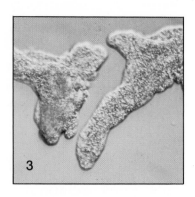

1

2

3

Ameba: A tiny animal found in pond water.

Hydra: A tiny animal found in lakes and ponds.

Life cycle: The series of stages in the life of an organism.

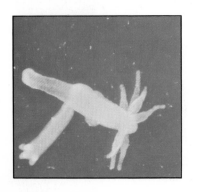

Look at the picture above. Picture 1 shows a very tiny animal found in pond water. It is called an **ameba** (ah-**mee**-bah). An *ameba* does not hatch from an egg. The ameba population grows in a different way. Each ameba reproduces by splitting in half. Pictures 2 and 3 show how an ameba does this. It takes an ameba about 1 hour to divide.

A tiny animal found in lakes and ponds is the **hydra** (**hi**-drah). *Hydras* live on the stems and leaves of water plants. They reproduce by budding. They form little buds on their bodies. Each bud grows into a new hydra. The picture in the margin shows how this happens.

Brine shrimp, amebas, and hydras are each born in different ways. Birth is the first step in the **life cycle**. All living things have *life cycles*.

There are three steps, or stages, in the life cycle of any living thing: (1) birth, (2) growth and development, and (3) death. They are shown on page 245. The length of a life cycle can be a few moments or a hundred years. Let's look at the life cycle of a butterfly.

The butterfly's life begins as an egg. The egg is laid by a female butterfly. The egg hatches

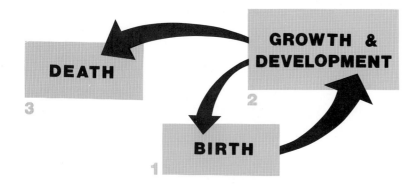

DEATH **3**

GROWTH & DEVELOPMENT **2**

BIRTH **1**

into a **larva** (**lar**-va). A *larva* is the young form of the organism. The larva of a butterfly is called a caterpillar. Caterpillars spend most of their lives looking for food. Many are eaten by birds. In time, the caterpillar stops growing. At this stage, it makes a chrysalis. It becomes a **pupa** (**pyoo-puh**). During the *pupa* stage, it changes into an adult. Its wings, mouth, and legs grow. It becomes a butterfly. Adult butterflies lay eggs. The life cycle begins again. At some time, the butterfly dies. Death is common to all living things. The organism might die of old age. Or it might be eaten by another organism. It might also die from disease. It might not be able to get enough food.

Larva: The young form in the life cycle of an organism.

Pupa: A stage in the life cycle of an insect in which the insect is not active.

ACTIVITY

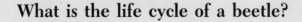

What is the life cycle of a beetle?

The beetle goes through the same stages of life as a butterfly.

A. Gather these materials: 2 or 3 mealworms, jar with a lid, bran, slice of potato or apple, and ruler.

B. Look at the mealworm. It is the larva stage of the beetle.

  1. Does the mealworm have legs?
  2. What does it eat?
  3. When did the larva stage end?
  4. What did it look like at this stage? What is this stage called?
  5. What are the stages of a beetle's life cycle?

Section Review

Main Ideas: An animal goes through stages during its lifetime. These stages are birth, growth and development, and death.

Questions: Answer in complete sentences.

1. Describe the life cycle of a sea turtle.
2. What are three ways animals reproduce?
3. What are some reasons that living things die?
4. Draw the life cycle of a butterfly. Label the stages.
5. What is the life cycle of a human?

Have you ever wondered what it would be like to be a tree? These are redwood trees. They are the largest and nearly the oldest living things on earth. Some redwoods are over 3,500 years old. They can be as high as 90 meters (300 feet). How are new redwood trees made? When you finish this section, you should be able to:

Plant Life Cycles

☐ **A.** Describe and give examples of plant life cycles.

☐ **B.** Compare life cycles of plants that make seeds with life cycles of plants that do not make seeds.

Like animals, plants have life cycles. Plant life cycles fall into two groups.

Group 1: Cycles of plants that make seeds
Group 2: Cycles of plants that do not make seeds

Some plants make their seeds in flowers. Other plants make their seeds in cones. The drawing shows the life cycle of a cone-bearing tree. What are the steps in the life cycle?

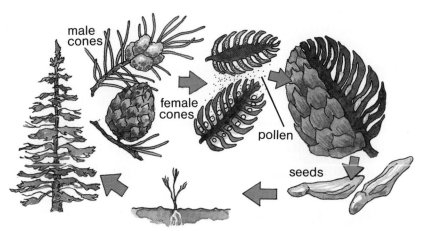

male cones

female cones

pollen

seeds

247

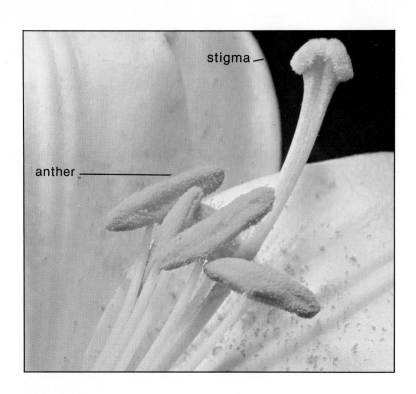

stigma

anther

Anther: The part of a flower that contains pollen.

Pollen: A yellow powder made by the anthers of a flower.

Stigma: The part of a flower that receives pollen.

Let's look at the life cycle of plants that make seeds from flowers. Many plants have flowers. Most flowers usually open in warm weather. Look at the flower shown above. One part of the flower is called the **anther** (**an**-thur). *Anthers* make a yellow powder called **pollen** (**pahl**-len). Another part of the flower is called the **stigma** (**stig**-mah). The *pollen* must move from the anther to the *stigma*. Then a seed can begin to form. Bees and other insects help move pollen. When insects walk on flowers, they catch pollen on their bodies. They carry the pollen from the anthers to the stigmas. Wind, water, birds, and other animals also carry pollen. Once pollen reaches a stigma, the flower is pollinated. Soon after, seeds begin to form. Most new plants grow from seeds.

Fruit trees are flowering plants. The fruits grow around the seeds. When fruits fall to the ground, their seeds stay in the soil. They can begin to grow into new fruit trees. In a few years, the new trees will flower. Seed formation starts all over again. The new trees add to the size of the fruit tree population. Plants must reproduce if a population is to survive.

Plants can also reproduce themselves in other ways. Some can make new plants if you put pieces of their roots, stems, or leaves in soil or water. Potatoes grow tiny buds called eyes. If you plant a piece of potato with the bud in it, a new potato plant will grow from the bud.

Strawberries reproduce by means of runners. A runner is a stem that grows on the ground. When the runner touches the ground, it takes root, and a new plant develops.

Fern: A simple, green, non-flowering plant.

Moss: A small green plant that grows on rocks.

Spores: Small, round objects found on ferns, which can grow and form new fern plants.

Plants that do not produce seeds reproduce in many other ways. **Ferns** and **mosses** reproduce without seeds.

Ferns and *mosses* reproduce by means of **spores**. These *spores* are much smaller than grains of salt. Look at the life cycle of the fern. Ferns are green plants with roots, stems, and leaves. Fern spores are produced in special places on the underside of leaves. They can fall to the ground. If a spore falls on wet ground, it will grow into a thin, green, heart-shaped plant. The adult fern grows from this tiny plant. Most ferns have life cycles that last for years.

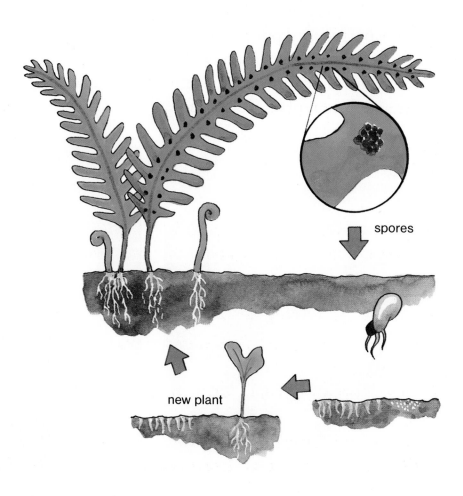

spores

new plant

ACTIVITY

How do mushrooms reproduce?

A. Gather these materials: a mushroom, a piece of light-colored paper, and a dish.

B. Break off the mushroom cap. Place it on the paper, underside down.

C. Cover the cap with the dish and leave it for one day.

D. Gently lift the dish and the cap.
   1. What do you see?
   2. Do mushrooms reproduce by spores or seeds?
   3. What conditions are necessary for new mushrooms to sprout?

Section Review

Main Ideas: Some plants make seeds and some do not make seeds. Plants can also reproduce if their stems, leaves, or roots are cut, and from runners.

Questions: Answer in complete sentences.

1. What are three ways in which plants can reproduce?
2. How does the plant shown in the margin reproduce?
3. Draw the life cycle of this plant.
4. In what way is a spore like a seed?

251

Organisms on the Move

The animal shown in this picture is a harp seal. It lives near the Arctic Circle. In the summer, it may swim over 700 miles north of the Arctic Circle. When winter comes, herds of seals move south. Why do you think harp seals move south in winter? When you finish this section, you should be able to:

☐ **A.** Explain why animals *migrate* to new places.

☐ **B.** Identify the *migration* routes of populations of animals.

☐ **C.** Explain how plant populations move to new places.

Most populations of animals live all their lives in the same place. Single organisms, like the black bear, might roam through a large forest looking for food. But most populations make their home in one place and stay there.

Populations of birds, insects, and mammals move each year. The movement of animals from place to place is called **migration** (my-**gray**-shun). Why do animals *migrate*?

Scientists think that animals migrate for several reasons. Many animals migrate in winter in order to reach warmer areas. Some animals migrate to find new sources of food. Others migrate to areas where they can raise their young under better conditions.

Migration: The movement of animals from place to place.

Bighorn sheep: An animal with curved horns that lives in the Rocky Mountains.

Monarch butterfly: A beautiful butterfly that migrates over long distances.

Harp seals migrate from north to south. Other animal populations migrate from east to west. This brings them to warmer places. Some populations migrate from high country to low country. For example, the **bighorn sheep** spends summers in the high meadows of the Rocky Mountains. In winter, the sheep move down to low country. They are moving away from the cold winters.

Another animal that migrates is the whale. Scientists think whales migrate mainly to reproduce. They spend their summers in cold polar seas. During the winter they travel to warm seas. There they give birth and raise their pups. When spring comes, they return to the polar waters.

Some insects also migrate. The **monarch butterfly** (**moh**-nark) migrates over 1,600 kilometers (1,000 miles). At the end of each summer, it flies south from Canada. Some *monarch butterflies* go to the Gulf of Mexico. Others go as far as Mexico. In the spring, they begin their migration north.

You may have seen migrating birds in flight. One kind of migrating population is the warbler, such as the yellow warbler. Warblers leave South America early in March. Ten days later, they reach Florida. Warblers return to the places where they nested the year before. Warblers can fly 50–80 kilometers (30–50 miles) per hour. But they stop to rest and eat on migration. So the warblers do not travel north that fast.

Plant populations can move to new places,

too. But plants move for other reasons. Plant migration is harder to see. It takes place slowly. When plants move into a place, it is called **succession** (suck-**seh**-shun).

Suppose a volcano erupts. It destroys all the plants and animals in an area. Only bare soil and rock are left. When the ground cools, *succession* starts. Wind or animals carry seeds into the area. The first seeds to come in and grow are grasses. Soon, seeds from other plants are carried in. A meadow of grasses and small plants forms. The next stage in succession is the arrival of cone-bearing plants. Pine trees start to grow. In the last stage, hardwood tree seeds arrive. Oak trees and maple trees start to grow. The movement of plants is very different from the movement of animals.

Succession: The movement of plant populations into an area.

255

ACTIVITY

What are some ways that seeds can travel?

A. Gather these materials: seeds, plastic cup, and water.

B. Look at each seed. Think of all the ways that a seed could travel. Drop it through the air. See if it floats in water. Try to stick it to your clothing.

    1. How does the shape of the seed affect the way it might travel?

    2. What parts of the seed help it travel?

    3. What are some ways that seeds travel?

    4. Why is it important for seeds to travel?

Section Review

Main Ideas: Plants and animals can move from place to place. Animal populations migrate to find food, to reproduce, or to reach a warmer place.

Questions: Answer in complete sentences.

1. Do bighorn sheep migrate for the same reason as whales? Explain your answer.
2. What is the migration route of the monarch butterfly?
3. How would a new population of plants reach a new place?
4. How does animal migration differ from the way plants move to new areas?

Population Explosions

The plant shown here is Kudzu. About 40 years ago, people brought this plant from Japan. Soon, Kudzu was growing wild. What might have caused this to happen? When you finish this section, you should be able to:

☐ **A.** Identify the cause of population explosions.

☐ **B.** Identify four things that cause the size of a population to change.

☐ **C.** Predict what may happen if a population gets too large.

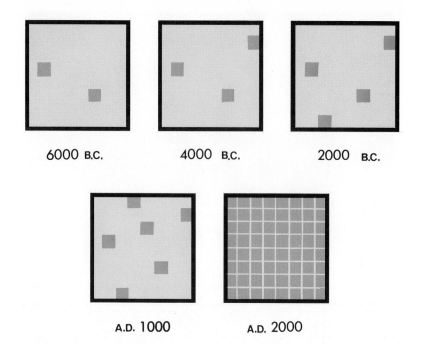

6000 B.C. 4000 B.C. 2000 B.C.

A.D. 1000 A.D. 2000

A population that grows very fast is said to be exploding. When this happens, there is not enough living space for the population. Look at the diagram above. It shows the number of humans on each square mile of the earth. In 6,000 B.C., there were only two people for each square mile. For thousands of years, the population did not change much. But between A.D. 1,000 and the present, the human population exploded. How many people are now living on a square mile?

The drawing shows some of the ways in which a population can change. Populations grow if there are more births than deaths. They also grow if more of their members move into the area. Populations decrease if more deaths occur than births. They also decrease if their members move out of the area.

Births

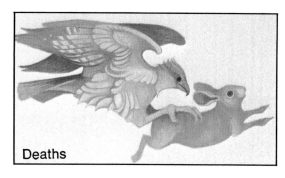

Deaths

Moving into a Place

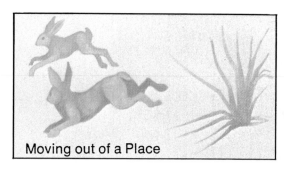

Moving out of a Place

Let's look more closely at the changes in a population. Suppose you have a jar of flies like the one shown. You put food in the jar. Because there is food, the population of flies grows. You can see this in the second jar in the picture. What happens to the population of flies after a longer time? There is still plenty of food. But the population has decreased. Many of the flies have died because of the buildup of wastes in the jar.

1   flies   food

2   flies

3   dead flies

In this case, there was too little space for the flies. If the jar were bigger, more flies could live in it. But even that space would soon fill up.

In the next case, a lack of food limited the size of a deer population. Deer populations can grow very fast. Not long ago, the population of deer in the Florida Everglades exploded. The Everglades is a swamp. One year it flooded. The plants that the deer eat were covered by water. There was less food for the deer. Many of the deer became weak and sick. The sick deer died. The population started to decrease.

Some populations eat other populations. The fox shown here eats lemmings. If there are few foxes, the lemming population grows. Suppose the foxes were killed by a disease. What do you think would happen to the lemming population? What do you think would happen to the lemming population if the foxes were healthy?

ACTIVITY

How does the amount of space affect the size of the population?

A. Gather these materials: 1 small, 1 medium, and 1 large jar; Daphnia; algae water; medicine dropper; and aged tap water.

B. Pour equal amounts of algae water into the 3 jars. Fill the jars with aged tap water.

C. Put the same number of Daphnia (about 10) in each jar. Put the jars in a place where there is plenty of light. It should not be too hot.

D. Each day for the next week, count the number of Daphnia in each jar.

   1. How did the populations change?

   2. What effect did the size of the jar have on the size of the population?

Section Review

Main Ideas: The size of a population depends on births, deaths, and the movement of members in and out of an area.

Questions: Answer in complete sentences.

1. What are two things that can cause a population to change size?

2. Do population explosions last forever? Why?

3. Is the human population getting bigger or smaller? What effect may this have?

CHAPTER REVIEW

Science Words: Think of a word for each blank. List the letters **a** through **m** on paper. Write the word next to each letter.

____**a**____ are small animals that live in salt water. You can find ___**b**___ in lakes and ponds.

The young form of an insect is called a ___**c**___. The next stage in an insect life cycle is the ___**d**___ stage.

The part of the flower that contains the pollen is the ___**e**___. The yellow powder made by the anther is called ___**f**___. The part of the flower that receives pollen is the ___**g**___. A simple green plant is a ___**h**___. A ___**i**___ is a small round object found on ferns.

The movement of animals from place to place is called ___**j**___. An animal with curved horns is the ___**k**___. The name of a butterfly that migrates is the ___**l**___. The movement of plants into an area is called ___**m**___.

Questions: Answer in complete sentences.

1. What are some ways that animals reproduce? Give at least two ways.
2. What are two ways that plants can reproduce?
3. Do plants migrate? Explain your answer.
4. Describe two ways that population size can be controlled.
5. Describe the life cycle of a fern. Draw a diagram with labels to show your answer.

CHAPTER 16

SURVIVAL AND CHANGE

You should be able to tell the animals from the plants shown here. You can also see the difference between the two animal populations. But what about one population? When you finish this section, you should be able to:

16-1.

Comparing Organisms

☐ **A.** Identify ways in which populations differ.

☐ **B.** Compare organisms of the same population.

Look at these pictures of frogs. Each frog is from a different population. Their colors are different. They are also different sizes. What are some ways in which they are alike?

Populations differ in many ways. List the ways in which you think animal populations can differ.

One way populations differ is in size.

A virus is very small. You need a very strong microscope to see it. It can show the virus enlarged 1,000 times. Four thousand viruses could fit on the period at the end of this sentence. Other populations are very large.

The largest animal is the blue whale. It is over 30 meters (100 feet) long. It weighs over 135,000 kilograms (150 tons). The largest plant is the Sequoia (seh-**kwoi**-ah) tree. It is over 90 meters (300 feet) tall.

Animals and plants are the sizes they are for a reason. Their size helps them survive. Giraffes can eat food that other animals cannot reach. They can also spot their enemies quickly. With their long necks, they can see over bushes and trees. Some monkeys can move quickly because they are so small. Their speed helps them get food and avoid enemies. Tall trees can get extra sunlight. They tower over smaller trees.

Differences can occur within a population. Look at the picture in the margin. It shows a population of land snails. Do you see any differences in size? Find the largest and the smallest snail. What are some other differences that you can see?

The pictures below show two different populations. What differences can you find among the living things in each population?

ACTIVITY

How large are the blue whale and the giant sequoia?

A. Get a meter stick. With another person, go to a large open area.

B. Take 10 average steps. Measure the distance you covered. Divide this distance by 10 to get the average length of one step.

C. The blue whale is 30 meters long and the giant sequoia is 90 meters long. Measure out these lengths in footsteps.

Section Review

Main Ideas: Members of different populations differ in size, shape, and color. Living things in the same population can also differ.

Questions: Answer in complete sentences.

1. List some ways that you differ from your mother and father.
2. What are three differences between a population of horses and a population of giraffes?
3. Find three differences in the population of leaves shown here.

16-2.

Changing Populations

Once upon a time, there was a population of elephants that lived on an island. At first, there were both large and small elephants. After a long time, however, the population was made up only of small elephants. What caused this change? When you finish this section, you should be able to:

☐ **A.** Explain how a population can change over time.

☐ **B.** Explain how *variations* can affect an organism's chance for survival.

You learned in the last section that there are differences among members of the same population. These differences are called **variations** (vare-ee-**ay**-shuns). *Variations* among organisms might include color, shape, or size.

Variations: Differences among members of the same population.

Variations can affect the survival of a population. Animals that survive can reproduce. The variations that helped them survive are passed on to their young. Let's look at the elephants on the island.

In this case, the island had too little food for all the elephants. The small elephants needed much less food than the large elephants. They were better able to live and reproduce. After a time, there were only small elephants living on the island.

Color can also affect survival. Suppose there are two colors of insects in a population. The two colors are green and red. The green insects are more likely to survive in a grassy place. They are not as easy to find as the red insects.

The picture on the left shows a hare. This hare can change its color. In the winter, its coat is white. In the summer, the color is brown. The second picture shows a hare from a different population. Its coat does not change. The color stays brown all year long. Which hare do you think would survive in the place shown in the two pictures?

ACTIVITY

How does an organism's color affect its chance to survive?

A. Gather these materials: red and green toothpicks.

B. Copy the chart shown. Pretend the toothpicks are insects. You will be a bird that eats insects.

C. When your teacher says "Go," cover 1 eye with your hand. Pick up toothpicks with your other hand. After 20 sec, your teacher will tell you to stop.

D. Record the number of red and green toothpicks you picked up.

E. Repeat steps C and D twice.

1. How many red toothpicks did you pick up? how many green toothpicks?

2. Which color toothpick could you see more easily?

3. Which color "insect" has the better chance to survive?

TOOTHPICKS	Red	Green
1		
2		
3		
Total		

Now let's look at what happened to the peppered moths in England. Peppered moths rest on tree trunks during the day. Birds feed on the peppered moths. Before factories were built in the 1850's, most tree trunks were light in color. Birds could see the dark-colored moths more easily than the light-colored moths. More of the dark moths were eaten. So more light-colored moths survived and reproduced. But what if the tree trunks changed color? The factories put black soot into the air. The soot settled on the bark of the trees. Many of the tree trunks became black. Soon, the number of dark moths increased. The birds did not see them. They blended in with the dark tree trunks. The birds ate more light-colored moths. The dark moths survived.

Section Review

Main Ideas: Populations can change over time. The change is due to variations among the members of the population. Changes in the environment can affect which members of a population survive.

Questions: Answer in complete sentences.

1. How can variations affect the chance that a living thing will survive? Give two examples.
2. Are all changes in the environment helpful to a population? Explain.
3. Look at the picture of the moths on this page. Explain how this population of moths changed over time.

16-3.

Disappearing Populations

Extinct: Describes a population that no longer exists.

A long time ago, dinosaurs roamed the land. There were many populations of dinosaurs. Dinosaurs lived on earth for over 160 million years. Then they were gone. Why did they die? When you finish this section, you should be able to:

☐ **A.** Identify populations that are *extinct* or in danger of becoming *extinct*.

☐ **B.** Explain what might cause the *extinction* of populations.

Organisms that no longer live on the earth are called **extinct** (ek-**stinkt**). Dinosaurs became *extinct* millions of years ago. On page 273 are pictures of two other extinct animals. Picture 1 shows a Labrador (**lab**-bra-dore) duck. Picture 2 shows a passenger pigeon. Both of these birds became extinct within the last 100 years.

More than 170 kinds of animals and 1,700 kinds of plants are now rare. Their numbers are decreasing. Someday, all of them may be gone. They are all in danger of becoming extinct. They are called **endangered** (en-**dane**-jurd) organisms. Can you name an organism that is in danger of becoming extinct?

The animals shown on page 274 are *endangered*. They are the bald eagle and grizzly bear. There are many reasons why living things become endangered. A change in the weather over a long time is one reason. Suppose warm temperatures caused the ice around the North and South Poles to melt. Some of the land would be flooded. Many plants and animals would not be

Endangered: Describes a population in danger of being extinct.

273

able to live. They would slowly die until none were left.

Human beings can endanger plants and animals. Many buffalo, beavers, and otters were killed. Their skins were used for clothes. As the human population grows, there is less room for animal and plant populations.

Endangered organisms are now protected. It is a crime to hunt them. Is it important to protect animals and plants? What difference does it make to other living things if an animal or plant becomes extinct? You will learn the answers to these questions in the next unit.

A change in the environment can also endanger organisms. The ivory-billed woodpecker feeds only on insects in dead trees. The forests where they lived were cut down. The old trees were replaced with new trees. This destroyed their food supply. The birds became extinct. Biologists asked that some old trees be allowed to remain. How would this help the ivory-billed woodpecker to survive?

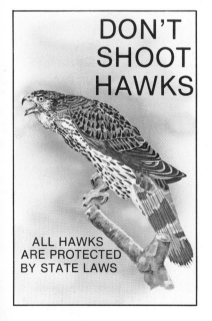

DON'T SHOOT HAWKS

ALL HAWKS ARE PROTECTED BY STATE LAWS

The chart below lists some of the extinct and endangered animals in North America.

Some Extinct North American Animals	
Passenger pigeon	Ivory-billed woodpecker
Heath hen	Carolina parakeet
Great auk	Labrador duck
Some Endangered North American Animals	
Whooping crane	Puma
American bald eagle	Musk ox
American crocodile	Trumpeter swan
California condor	Key deer
Grizzly bear	Blue whale

Some whales are almost extinct. Too many of them have been killed. Laws have been passed to limit the number of whales that can be killed. But a few countries like Russia and Japan still hunt whales.

Why did the dinosaurs become extinct? Some scientists think that too many dinosaur eggs might have been eaten. Others think that the climate became cooler. The dinosaurs were used to a warm climate and died. Dinosaur eggs with very thin shells have been found. This may mean that some disease was killing the animals. Or a giant asteroid may have hit the earth. This may have caused a giant dust cloud. The cloud would have blocked light from the sun. Plants need sunlight. If the plants they ate died, many dinosaurs also would have died. Scientists have tried to explain why the dinosaurs became extinct. No one knows for sure why they died. What do you think happened?

Section Review

Main Ideas: Many populations have become extinct. Others are in danger of becoming extinct. Endangered populations are protected by law.

Questions: Answer in complete sentences.

1. What is the difference between an extinct and an endangered population?
2. Name an endangered population. Why is it endangered?
3. What might have happened to the dinosaurs?
4. Why should we stop living things from becoming extinct?

CHAPTER REVIEW

Science Words: Match the terms in column A with the definitions in column B.

Column A	Column B
1. Variations	a. A population that might become extinct
2. Extinct	b. Differences among members of the same population
3. Endangered	c. A population that does not exist anymore

Questions: Answer in complete sentences.

1. What are some differences in the populations shown on page 263?
2. What are some differences you might see in a population of trees?
3. The peppered moths in England changed from light to dark. How did this happen?
4. Giraffes eat leaves. Which type of giraffe shown in the picture would have a better chance of surviving if the trees were tall? Explain.
5. Squirrels are very fast. How does a squirrel's speed help it survive?
6. Each of these living things is endangered: bald eagle, humpback whale, California condor, grizzly bear. Choose one. Tell why it is endangered.
7. What are some things that you could do to help stop the extinction of plants or animals?

INVESTIGATING

What populations are near your school?

A. Gather these materials: magnifying glass and a stick to dig with.

B. Copy the chart shown below.

C. When you see a living thing, make a mark for that population.

D. You will search in an area 10 m × 10 m. Your teacher will take you to the spot.

E. Look for living things that are shown on the chart. Use the stick to dig around rocks and roots. Be careful not to destroy the area.

    1. What is the largest population?

    2. What is the smallest population?

    3. Are there more animal populations or plant populations?

    4. Do you think the populations will be the same next month? in 6 months? Why?

ORGANISM

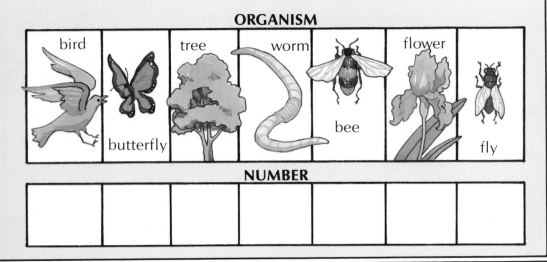

bird	butterfly	tree	worm	bee	flower	fly

NUMBER

CAREERS

Horticulturist ▶

Horticulturists (hore-ti-**kull**-chur-ists) are people who work with plants. They work in greenhouses and gardens. They try to improve the plants they grow.

A visit to a greenhouse will help you understand what horticulturists do. They must know about soils and fertilizers.

A horticulturist needs to go to school after high school.

◀ Entomologist

Entomologists (en-toe-**moll**-eh-jists) study insect populations. They may study termites, cockroaches, bees, ants, or butterflies.

Some entomologists try to find ways to control harmful insects. Others do research. They work in universities. Entomologists need a college degree in biology. Many people study and collect insects as a hobby.

Animal and Plant Communities

UNIT 6

CHAPTER 17

ENERGY FOR LIVING

17-1.

Communities

Can populations live alone? Scientists know of no living population that exists apart from other living things. How do different populations get along with each other? When you finish this section, you should be able to:

☐ **A.** Identify how populations relate to each other in a *community*.

☐ **B.** Describe the features of a *community*.

☐ **C.** Describe life in different *habitats*.

There are very few places on earth where living things cannot be found. Most living things live near other living things. All the plants and animals that live in the same area are called a **community** (kuh-**mew**-nit-tee). There are many kinds of *communities*. Communities of clams, crabs, snails, and algae live in the ocean. Grass and prairie dogs (**prayr**-rhee) are community members of the grasslands, or prairie.

Camels and cacti are community members of a desert. Communities of frogs, turtles, pond lilies, and fish live in ponds.

Each member of a community lives in a certain place. That place is called a **habitat** (ha-bi-tat). There are many *habitats* within a com-

Community: All the plants and animals that live in the same area.

Habitat: The place where a thing lives within a community.

munity. In a forest, the soil is the habitat of earthworms and small insects. A salamander's habitat may be an old log. Deer and birds live among the trees and shrubs.

In a pond, cattails grow in the moist soil and mud at the water's edge. Pond lilies are found floating on top of the water. The habitat of frogs, tadpoles, small fish, and insects may be under rocks, under plants, or in the mud. Can you think of any other habitats?

Populations need each other in order to survive. In any community, plants, animals, and protists work with each other. There are a number of ways in which living things work with each other. Let's look at a few.

A dead log is a habitat. A population of termites lives in the log. The termites eat the wood. However, they cannot digest the wood. Protists live inside the termites. The protists

digest the wood for the termites. This is an example of how organisms work with each other. The dead log is a home for the termites. The termites eat the wood. The protists have a "home" inside the termites. The protists help the termites. They digest their food. The way these organisms depend on each other is called **mutualism** (**mew-chew-uhl-ism**). *Mutualism* is a relationship in which two or more organisms help each other. What are some other examples of mutualism?

Mutualism: A relationship in which two or more organisms live together and help each other.

ACTIVITY

What are the features of a habitat?

A. Gather these materials: 2-gallon glass or plastic container with lid, potting soil, seeds (grass, radish, wheat), and water sprinkler.

B. Put soil in the container until it is about 1/3 full.

C. Spread all the seeds in the container.

D. Water the seeds and soil.

E. Put your habitat in a sunny, warm place.

F. Make a drawing of your habitat each day. Show the changes that are taking place.

   1. What would happen to the seeds if there were no soil?

   2. Which seeds do you think will grow first?

   3. Do you think all the seeds will grow? Explain.

Parasite: An organism that feeds off another living organism.

Sometimes, only one organism is helped. The other is not. One organism takes something from the other or harms it. Suppose you are bitten by a mosquito. The mosquito is a blood sucker. It uses your blood as food. The mosquito is helped. But you may be harmed. The mosquito gets food. You may get a disease caused by protists the mosquito carries. Biologists call the mosquito a **parasite** (**pare**-uh-sight). A *parasite* is a living thing that feeds off other living things.

There are many other ways in which organisms in a community interact. You will learn about these ways in other sections of this unit. The thing to keep in mind is this: Organisms in communities depend on each other. No organism lives alone.

Section Review

Main Ideas: Plants, animals, and protists live together in communities. Each member of a community has its own habitat. The organisms in a community work with each other.

Questions: Answer in complete sentences.

1. Describe three different communities.
2. How do the habitats of a tadpole and an earthworm differ?
3. Give an example of the way organisms work together in a community. Explain.
4. What do you think would happen to the organisms in a pond if the pond dried up?

When you see a green plant, you really are look-ing at a factory. Factories make things. Some factories make cars, airplanes, clothes, or chairs. If green plants are like factories, what do they make? When you finish this section, you should be able to:

☐ **A.** Identify the things green plants need to make food.

☐ **B.** Explain what is meant by *photosyn-thesis.*

Green plants are very different from animals. When your hand is in sunshine, it gets warm. Nothing else happens. A plant does not have hands. But it does have leaves. Leaves that are in sunshine can make food. Green plants are the only living things on earth that can make food.

Look at these leaves from different plants. Plants make their food in the leaves.

The picture below shows what a green leaf looks like under a microscope. The tiny box-like objects are **cells**. Food for the plant is made in the *cells*. If you look closely, you can see tiny green parts inside each cell. This material is called **chlorophyll** (**klore**-oh-fill). The *chlorophyll* gives the plant its green color. Chlorophyll makes it possible for green plants to make food.

The way green plants make food is called **photosynthesis** (fo-to-**sin**-thuh-sis). *Photo-* means "light," and *-synthesis* means "to put together." In *photosynthesis*, a plant uses sunlight. The

Cell: The smallest unit of which all organisms are made.

Chlorophyll: The green material inside a green plant cell.

Photosynthesis: The way green plants make food.

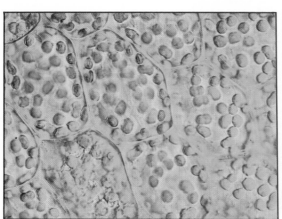

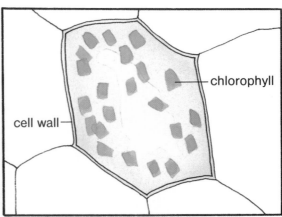

sunlight helps it put together air, water, and minerals to make food. The food that plants make is sugar. Let's look more closely at how green plants combine these things to make food.

The picture below shows the parts of a green plant. The materials for photosynthesis must get to the leaf cells. Water and minerals come from the soil. They enter the plant through its roots. Tiny root hairs let water and minerals come into the root. The water and minerals travel from the roots to the stems or trunk of the plant. Inside the stems are thin tubes. The water is carried up the tubes to the branches. Then it is carried to the leaves.

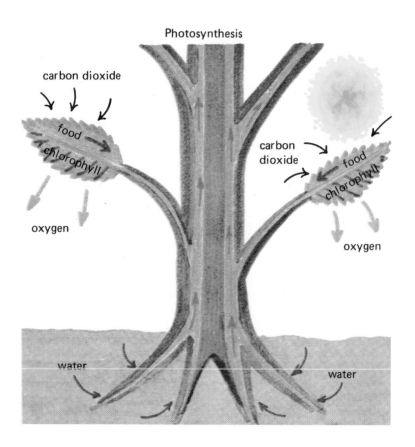

Photosynthesis

carbon dioxide

food

chlorophyll

oxygen

carbon dioxide

food

chlorophyll

oxygen

water

water

Carbon dioxide: A gas that is needed by green plants to make food.

Oxygen: A gas that is given off by the leaves of green plants.

The picture on page 289 also shows that green plants need light and a gas called **carbon dioxide** (**kar**-bon die-**ox**-ide). The *carbon dioxide* enters the plant through holes on the bottom of the leaves. The sunlight provides energy. The plant uses the energy to combine carbon dioxide and water to make sugar. During photosynthesis, the leaves give off a gas. This gas is called **oxygen** (**ox**-i-jen).

ACTIVITY

How do water and minerals get to the leaves of a green plant?

A. Gather these materials: celery stalks and leaves, 2 small jars, water, food coloring, knife, plastic bag, and rubber band.

B. Fill the 2 jars with water. Add a small amount of food coloring.

C. Cover the stem of 1 piece of celery with a plastic bag. Fasten it with a rubber band.

D. Put the covered stalk and uncovered stalk of celery in separate jars. Leave them in the jars for at least 1 hour.

E. Look at the 2 celery stalks.

    1. What do you see?

F. Cut the 2 stalks in half.

    2. How do you know that water moves up the stalk?

    3. How do you think the water gets to the leaves of a plant? Draw a picture to show how this happens.

Section Review

Main Ideas: The equation below shows how plants make their own food.

carbon dioxide + water + light + chlorophyll = sugar (food) + oxygen

Questions: Answer in complete sentences.

1. How is the cell of a green plant like a factory?
2. A plant needs water. What other things does it need to make its food?
3. Where do the things green plants need to make food come from?
4. How does a green plant make food?

People in Science

Rachel Carson

Rachel Carson was an ecologist. This means she studied the way plants and animals live together in their environments. In 1962, she wrote a book called *Silent Spring.* In her book, Carson said that people were poisoning the earth. We were using dangerous chemicals to kill insects that harm plants. These chemicals could also cause many plants to die. Soon there would be less food for animals. There would be no spring season on earth. Because of her book and others like it, laws were passed to protect the earth. But still more needs to be done to save plants and animals.

17-3.

Using Oxygen

The deer shown here is eating food that was made by plants. But it is also breathing air. The deer needs the oxygen in the air. What part of the biosphere makes the oxygen that animals need? When you finish this section, you should be able to:

☐ **A.** Identify the ways in which plants and animals depend upon each other for their food and energy.

☐ **B.** Explain what is meant by the *oxygen-carbon dioxide cycle*.

You learned in the last section that plants need light, water, minerals, and carbon dioxide. This is one way living things need non-living things. In the biosphere, living things also need each other. Animals need plants. Plants need animals. Let's find out why.

Plants need carbon dioxide for photosynthesis to take place. Carbon dioxide is a gas found in the air. It enters the plant through tiny holes in the bottom of the leaf. The picture below shows these tiny holes.

You learned that during photosynthesis, plants make sugar. Sugar is a source of energy. It is used by the leaves and the rest of the plant to grow. During photosynthesis, the plant also makes oxygen. This oxygen is given off by the plant. It goes into the air. Most of the oxygen in the air is made by green plants.

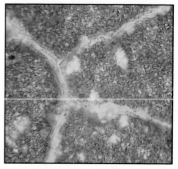

The two animals shown here need oxygen to live. They get the oxygen from the air by breathing. Oxygen helps animals get energy from the food they eat. Animals give off carbon dioxide when they exhale. The carbon dioxide goes into the atmosphere.

The diagram below shows how these gases move through the biosphere. They move back and forth between green plants and animals.

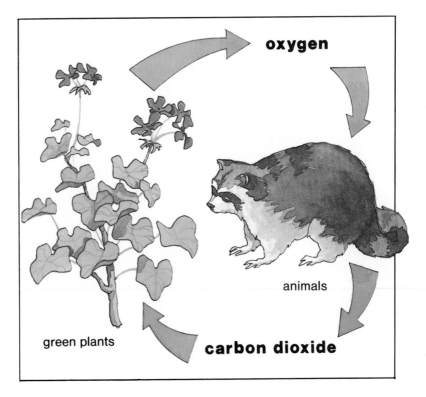

oxygen

animals

green plants

carbon dioxide

The chart on the bottom of page 294 looks like a circle. It shows a cycle. It is called the **oxygen-carbon dioxide cycle**. In this cycle, plants, animals, and the two gases interact. Oxygen and carbon dioxide are cycled through plants , animals, and the air. The biosphere and the atmosphere take part in the *oxygen-carbon dioxide cycle*.

This cycle also takes place in ponds and in the oceans. The picture below shows a pond community. Green plants and animals live together in the pond. The oxygen the animals need is in the water. The animals take the water into their bodies. They remove the oxygen and use it. Plants use the carbon dioxide in the water. The gases are cycled through the water just as they are cycled through the air.

Oxygen–carbon dioxide cycle: The way in which these gases interact with plants, animals, and the air.

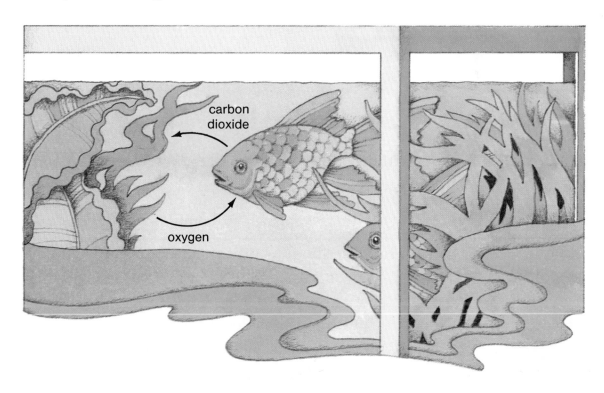

carbon dioxide

oxygen

ACTIVITY

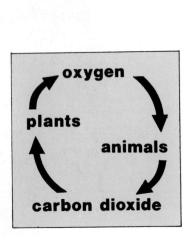

What gas do animals give off?

A. Gather these materials: 2 small bottles with day-old tap water, a straw, and a bottle of blue BTB.

B. Fill both bottles 3/4 full of day-old tap water. Add blue BTB until the bottles are full. Blue BTB turns yellow if carbon dioxide is in the water.

C. Put a straw into 1 of the bottles. Exhale through the straw.
   1. What is the color of the water in each bottle?
   2. What do you think happened in each bottle?
   3. What gas do humans and other animals give off?

Section Review

Main Ideas: The chart below shows the oxygen-carbon dioxide cycle.

Questions: Answer in complete sentences.

1. Describe how oxygen and carbon dioxide interact with a frog and a lily pad.
2. What is the source of these gases in the air: (a) carbon dioxide, (b) oxygen?
3. What would happen to animals if all the green plants died?

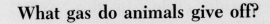

oxygen

plants

animals

carbon dioxide

CHAPTER REVIEW

Science Words: Think of a word for each blank. List the letters **a** through **j** on paper. Write the word next to each letter.

Plants and animals live together in a ___**a**___. Each organism lives in a home called a ___**b**___.

Two organisms of a different kind may live together and help each other. This is called ___**c**___. An organism that feeds on other organisms is called a ___**d**___.

The process by which green plants make food is called ___**e**___. Light is used by the green material found in the ___**f**___ of the plant's leaves. The green material is called ___**g**___.

The gas needed by plants to make food is called ___**h**___. The gas given off by green plants is called ___**i**___. Plants and animals pass these two gases back and forth in the ___**j**___.

Questions: Answer in complete sentences.

1. How do a habitat and a community differ? Name one of each.
2. What is one example of mutualism?
3. Look at the picture on page 282. Name the two organisms shown. What is the habitat of each?
4. What is the makeup of the community you live in? Describe your habitat.
5. In a drawing, show how green plants make food. Label light, chlorophyll, carbon dioxide, oxygen, and sugar.

CHAPTER 18

THE FOOD CYCLE

18-1.

Food Makers

Every living thing in a community needs food to live. There must be a source of this food. There are many different sources of food. In every community, there are food makers. When you finish this section, you should be able to:

☐ **A.** Identify the *producers* of food in a community.

☐ **B.** Identify where food is stored in plants.

Some of the food made by a plant is used to keep the plant alive. The rest is stored in different parts of the plant. The stems, roots, and leaves are all places where food can be stored. When you eat carrots, you are eating the roots of a plant. Broccoli is mostly the stem of a plant. Spinach and lettuce are leaves of plants. Peanuts are seeds. A fruit is food that has seeds.

Producers: Living things that are able to make their own food.

Cattails: A tall plant that lives in the water.

Water lily: A plant with large, flat, floating leaves with a flower in the center.

Because green plants can make, or produce, their own food, they are called **producers** (pro-**doo**-sirs).

Let's compare *producers* found in a water community with those in a forest community. The producers in a water community are green plants. There are four types of plants in this community: (1) Some plants have roots such as the **cattails** (**kat**-tails) shown in the picture below. (2) There are plants that live on the surface of the water. One of these is the **water lily**. (3) There are many plants that live under the water. (4) There are also tiny floating plants, such as algae. In water communities, algae are very important producers of food.

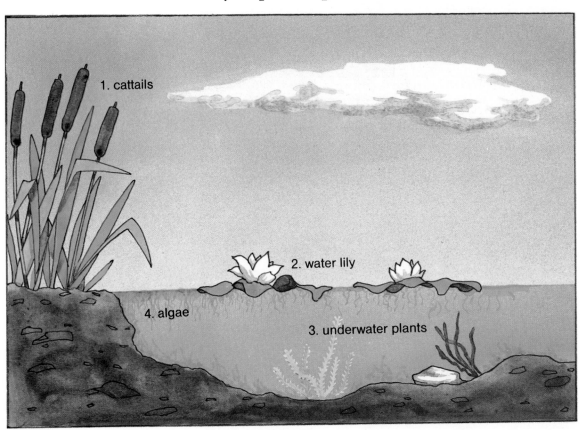

1. cattails
2. water lily
3. underwater plants
4. algae

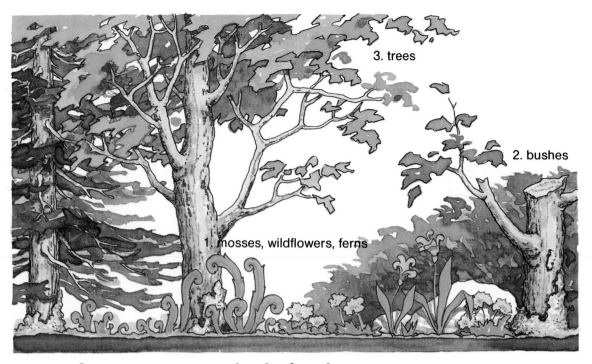

3. trees

2. bushes

1. mosses, wildflowers, ferns

In a forest community, the food makers are also the green plants. Look at the picture above. The food makers are shown in different places. There are three types of food makers in the forest. They are found in layers. (1) The bottom, or ground, layer contains mosses, wildflowers, and ferns. (2) The next layer is made of bushes and other low-growing plants. (3) The last layer is the trees. The trees and plants have roots, which grow into the ground.

The food made is in the form of sugar and **starch**. It is stored in different parts of the plants. It becomes a source of energy for living things that eat the plants. When you eat a plant food, you are eating sugar and *starch*. Foods that contain starch include bread, spaghetti, potatoes, rice, and cereals. Sugars are found in fruit, corn, and nuts.

Starch: A type of food made by green plants.

ACTIVITY

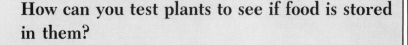

How can you test plants to see if food is stored in them?

A. Gather these materials: iodine solution, eye dropper, foods (cornstarch, rice, carrots, cereal, potato), paper towel, and a knife.

B. Put a small piece of potato on a paper towel.

C. Place 1 drop of iodine solution on the potato.

   1. What happens to the color of the iodine?

   2. Does this food contain stored starch?

      3. Which foods contain stored starch?

      4. What other foods do you think contain stored starch?

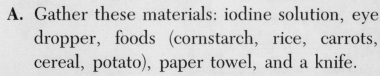

Section Review

Main Ideas: Green plants are called producers. They make their own food. Some food is used to keep the plant alive. Extra food is stored as starch or sugar in the stems, roots, and seeds.

Questions: Answer in complete sentences.

1. Which of the following are food makers: (a) corn, (b) horse, (c) lettuce, (d) turnip, or (e) sheep?

2. Where does a plant store food?

3. What are three types of producers in a water community?

4. What are some foods that you eat that contain starch?

Both of the animals shown here eat plants. The blue whale eats tiny organisms that float in seawater. It swims with its mouth open. It drinks gallons of water filled with the small organisms. The grasshopper eats plants that grow on the land. Neither the whale nor the grasshopper makes its own food. Each has to get its food by eating other organisms. In what ways are you like the whale and the grasshopper? When you finish this section, you should be able to:

☐ **A.** Identify organisms in a community that are *consumers*.

☐ **B.** Compare *consumers* based on the type of food they eat.

☐ **C.** Identify examples of *consumers* in a water and a land community.

Human beings and other animals are not producers. They are not green plants, which can make their own food. They get their food by eating plants and animals. The lettuce, sandwich bread, orange juice, and pear that the boy is eating come from producers. Lettuce is the leaf of a lettuce plant. Pears are fruit from pear trees. Orange juice comes from oranges, which are also fruit. Bread is made from wheat.

The ham on his sandwich does not come from a producer. Ham is meat. Meat comes from animals.

A living thing that eats plants and/or animals is called a **consumer** (kon-**sue**-mer). Human beings and all other animals are *consumers*. Some consumers eat only plants. These animals are called **herbivores** (**err**-bih-vors), or plant eaters. *Herb-* comes from a Latin word that means "grass." The suffix *-vore* comes from a Latin word that means "to eat." Cattle, deer, squirrels, mice, rabbits, grasshoppers, and butterflies are *herbivores*.

Consumer: A living thing that eats plants and/or animals.

Herbivore: A consumer that eats only plants.

Carnivore: A consumer that eats only animals.

Omnivore: A consumer that eats both plants and animals.

Some consumers feed only on other animals. These consumers are called **carnivores** (**kar**-nih-vors), or meat eaters. *Carni-* comes from a Latin word that means "meat." Foxes, coyotes, wolves, lions, seals, and frogs are *carnivores*.

Human beings eat both plants and animals. We are called **omnivores** (**om**-nih-vors). *Omni-* means "all." Bears, chickens, and some turtles are also *omnivores*.

Herbivores, carnivores, and omnivores are found in both water and forest communities.

A pond is a good place to find out about consumers in a water community. Look at the picture below. This is the same water community you saw on page 300. Let's compare the consumers in the pond. You cannot see the smallest consumers. The photo in the margin shows the tiny consumers magnified. These are tiny floating animals. They eat tiny floating plants. Since they eat plants, they are herbivores. You can also see other herbivores. Some insects, snails, and tadpoles are pond herbivores. Where are they found in the pond?

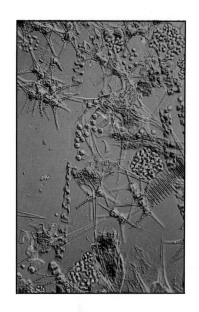

There are also many carnivores in the pond community. The crane is a water bird that eats fish. Can you find other carnivores?

ACTIVITY

How do producers and consumers interact in a community?

A. Gather these materials: 2 sheets of construction paper, small cards with the name of an organism, and glue.

B. Make a community chart on each sheet of construction paper. Label 1 chart "Forest Community." Label the other "Water Community."

C. Start with the cards labeled "Forest Community." Divide the cards into producers and consumers.

    1. What are some producers in the forest?

    2. What are some of the consumers?

D. Glue the cards on the community chart to show how the organisms interact.

    3. How do the consumers at the top of the chart get food from the producers at the bottom?

E. Repeat steps C and D with the cards labeled "Water Community."

    4. How is the water community like the forest community?

Look at the drawing of a forest community. This is the same forest that you saw on page 301. You can see many herbivores. The insects and small rodents, such as squirrels and chipmunks, are herbivores. Some of the larger animals are

also herbivores. Which of these are shown in the forest scene?

There are many carnivores. The owl eats insects. It also eats chipmunks and other small animals. Some of the birds eat worms and small insects. They also eat seeds. What kind of consumer is this type of bird?

The diagram on page 308 shows the three groups in a community. The plants are on the bottom. They produce the food for the groups above them. Next, we find the plant eaters. They become food for the animal eaters. This is a community chart. A community chart shows how energy moves through a community. The plants produce food by photosynthesis. This food energy is then used by the consumers.

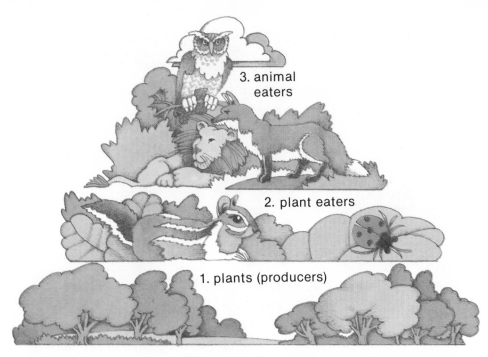

3. animal eaters

2. plant eaters

1. plants (producers)

Community Chart

Section Review

Main Ideas: Living things are either producers or consumers. Consumers can be either herbivores, carnivores, or omnivores.

Questions: Answer in complete sentences.

1. How is a producer different from a consumer? Name one of each.
2. What is the difference between a herbivore and a carnivore? Name one of each.
3. List the foods you ate for dinner last night. Which foods came from producers? Which came from consumers?
4. Draw a community chart for the pond shown on page 305. Identify one producer, one herbivore, and one carnivore. List them on the chart.

The Decomposers

Would you make a sandwich with this bread? The bread was once fresh. It was left out in the air. Now there is a black mold growing on it. What caused this to happen? When you finish this section, you should be able to:

☐ **A.** Explain what happens to organisms that die in a community.

☐ **B.** Identify organisms that feed on dead animals and plants.

☐ **C.** Explain how *decomposers* interact with producers and consumers.

Producers and consumers do not live forever. When they die, they begin to change. They begin to rot. We say that they **decay** (de-**kay**). Tiny organisms in every community make things *decay*. These living things get their food from wastes and dead organisms. They are called **decomposers** (de-kom-**poz**-ers). The prefix *de-*

Decay: To become rotten.

Decomposer: A living thing that feeds on wastes and dead organisms.

means "to undo." *Compose* means "to put together." So *decompose* means "to undo what is put together," or to break down. When decomposers feed on dead organisms, they make them decay by breaking them down.

Molds, yeasts, and bacteria (bak-**tear**-rhee-ah) are decomposers. They are very tiny non-green plants. They have no chlorophyll. They cannot carry on photosynthesis to make their own food.

Picture 1 is an enlarged photograph of mold growing on a slice of bread. Molds grow best where it is moist, warm, and dark. The bread shown on page 309 is decaying. A mold is living and growing on it.

Picture 2 shows bacteria. It would take 25,000 bacteria to fit on a line 2.5 centimeters (1 inch) long. Over 2,000 kinds of bacteria are known today. Some are harmful. Some are not. They are found in the air, on our skin, and inside our bodies. Certain bacteria grow only on plants. Others feed on animals. Still others grow only on humans. They all need food to survive.

Yeasts are shown in picture 3. Yeasts need sugar to live. They can live on the sugar in fresh fruit.

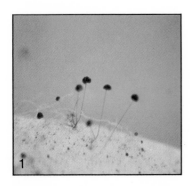

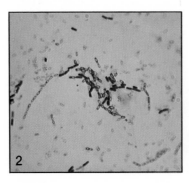

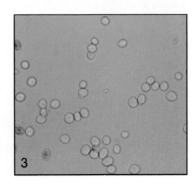

ACTIVITY

How do decomposers decay organisms?

A. Gather these materials: 2 banana slices, 2 plastic sandwich bags, and yeast.

B. Put a slice of banana inside each of the 2 plastic bags. Sprinkle some yeast on 1 slice.

C. Close both bags. Write "yeast" on the bag with the yeast.

D. Look at both bags each day for 5 days.
   1. What difference do you see between the banana with yeast and the one without yeast?
   2. How is the banana with yeast changing?
   3. What is causing this change?

Decomposers in a community are important. They break down and decay dead organisms. As they decay, these organisms give off materials. These materials go back into the community. Yeast, molds, and bacteria give off carbon dioxide as they feed on wastes and dead organisms. The carbon dioxide is then used by green plants to make food.

The diagram on page 312 shows decomposers, producers, and consumers in a community. This is another example of a cycle. (1) The green plants make food for animals. (2) The animals

materials return to soil and are used by plants to make new food

animal eats plants

yeast

bacteria

mold

decomposers breakdown dead organisms

both plants and animals die

and plants die. (3) The decomposers "eat" the dead organisms. The dead organisms are broken down. (4) The decomposers put materials back into the air and soil. These materials are used by green plants to make food.

Section Review

Main Ideas: Decomposers get their food from wastes and dead organisms. When organisms decay, the soil, air, and water get the materials needed to help keep other living things alive.

Questions: Answer in complete sentences.

1. What are three organisms that feed on dead organisms?
2. Suppose a squirrel dies in the forest. What will happen to its body after a few days?
3. Show in a diagram how trees, a squirrel, and bacteria are related in a community.

CHAPTER REVIEW

Science Words: Match the terms in column A with the definitions in column B.

Column A

1. Producer
2. Cattails
3. Water lily
4. Starch
5. Consumer
6. Herbivore
7. Carnivore
8. Omnivore
9. Decay
10. Decomposer

Column B

a. A living thing that feeds on wastes
b. Living things that make their own food
c. To become rotten
d. A tall plant that lives in the water
e. A consumer that eats plants and animals
f. A consumer that eats only plants
g. A plant with flat, floating leaves
h. A food made by green plants
i. A consumer that eats only animals
j. A living thing that eats plants or animals

Questions: Answer in complete sentences.

1. Make a list of the foods you ate today. Label the foods that are producers "P." Label the consumers "C."
2. Name three producers in a forest community.
3. Where do plants store food?
4. What kinds of consumers are these animals: (a) deer, (b) owl, (c) grasshopper, and (d) human?

CHAPTER 19

THE WEB OF LIFE

19-1.
Food Chains

Hawks have very good eyes. They can see small animals from great heights. Hawks eat mice, snakes, and small birds. How are these consumers related? When you finish this section, you should be able to:

☐ **A.** Explain what a *food chain* in a community is.

☐ **B.** Describe a *food chain* in a community.

The path that food travels in a community is called a **food chain**. A *food chain* describes how living things depend on each other for food. Most food chains include a green plant, a plant eater, and one or more animal eaters.

Each food chain leads to an animal that is not eaten by other animals. But this is not where a food chain ends. Members of food chains die. Then they become food for decomposers. The decomposers break down wastes and dead organisms all along the chain. The broken-down materials go back into the soil and water. They are used by producers to make more food.

To show a food chain, list each organism. Then draw an arrow from the food to the consumer. Look at the picture. It shows a food chain for a bird, a grasshopper, and grass.

Food chain: The path that food travels through a community.

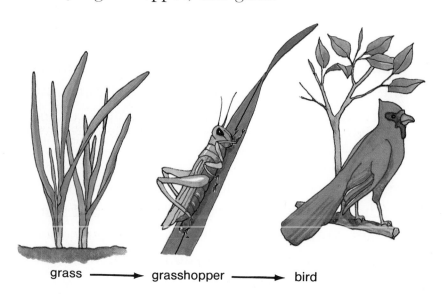

grass ⟶ grasshopper ⟶ bird

ACTIVITY

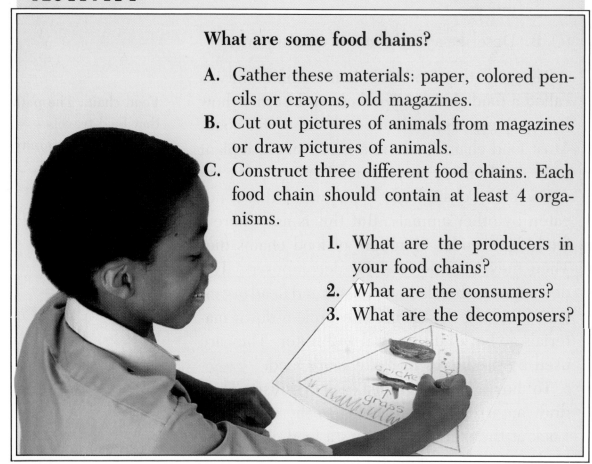

What are some food chains?

A. Gather these materials: paper, colored pencils or crayons, old magazines.

B. Cut out pictures of animals from magazines or draw pictures of animals.

C. Construct three different food chains. Each food chain should contain at least 4 organisms.

    1. What are the producers in your food chains?

    2. What are the consumers?

    3. What are the decomposers?

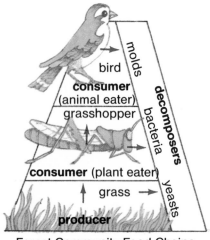

Forest Community Food Chains

Organisms exchange energy through a food chain. We can show this food chain on a community chart. The energy starts at the bottom of the chart. Energy comes from food. Plants produce food for the grasshopper. The grasshopper becomes food for the birds. As organisms die, they become food for the decomposers.

The drawing shows organisms in a forest. The arrows drawn from each organism show food chains. What are some of the food chains in the forest?

bird eating worm

squirrel eating acorns

deer eating grass

Section Review

Main Ideas: A food chain describes how energy travels in a community. Most food chains include a green plant, a plant eater, and one or more animal eaters. Decomposers are also part of a food chain.

Questions: Answer in complete sentences.

1. Draw a food chain for a leaf, a caterpillar, and a bird.
2. Draw a community chart. Describe a food chain in the community chart for the leaf, the caterpillar, and the bird.
3. What other organisms should be added to the community chart in question 2?
4. List one food you ate today. Draw a food chain that includes you, the food you ate, and the source of food for the food you ate.

19-2.

Connected Food Chains

The picture shows what scientists think life was like when the dinosaurs were alive. Some dinosaurs were herbivores. Others were carnivores. What are some food chains that might have existed at that time? Did the food chains connect with each other? When you finish this section, you should be able to:

☐ **A.** Trace a *food web* through a community.

☐ **B.** Tell the difference between a *predator* and a *prey.*

Suppose you are bitten by a mosquito. A friend says, "You are now a source of food for a hawk." Would you believe your friend?

Suppose the mosquito is eaten by a dragonfly. Then a frog eats the dragonfly. The frog is eaten by a snake. A hawk swoops down and eats the snake. The food the mosquito took from you ends up as food for the hawk! Can you write the chain of events from the mosquito to the hawk?

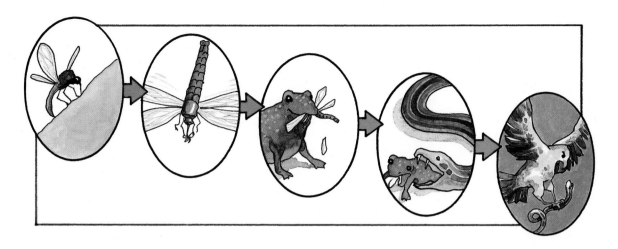

Most communities have many food chains. Most animals in a food chain eat more than one kind of food. Because of this, food chains often connect or overlap. Picture the food chains for seeds, a mouse, a snake, and a hawk. The food chains would look like the drawing. The mouse is food for both the hawk and the snake. The snake is food for the hawk. Food chains that overlap are called **food webs**. Sometimes, just two food chains overlap to form a *food web*. A food web may also have many overlapping food chains.

Here is a food web that you might find in many communities. There are many food chains that overlap. One food chain is tree→beetle→ bird→cat. Another food chain that overlaps the first one is tree→grasshopper→bird→cat. What is a third food chain in the web shown?

The food web on page 321 shows that squirrels eat nuts from the tree. Grasshoppers and beetles eat the tree's leaves. Bats and birds eat beetles and grasshoppers. Cats eat squirrels and birds. Five different food chains are involved.

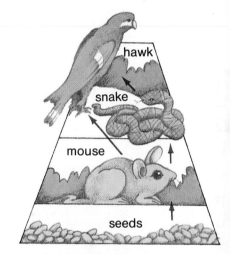

Food web: Food chains that connect or overlap.

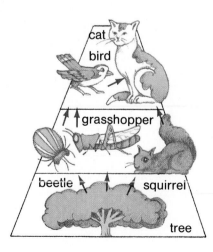

ACTIVITY

What is the food web in a forest community?

A. Gather these materials: board, pushpins, sheet of paper, 1 m of yarn, and scissors.

B. Put the paper on the board. Use the pins to make the paper flat on the board.

C. Next to each pin, write 1 of these names: oak tree, corn, caterpillar, hawk, owl, sparrow, squirrel, bacteria, decaying log.

D. Tie the end of the yarn around the pin next to "hawk." Which organism on the board might the hawk eat for food? If you think "squirrel," tie the string around "squirrel." You have a food chain between these two. Draw an arrow from "squirrel" to "hawk." This shows the direction in which the food travels.

E. Tie the string around the organism that is food for the squirrel.

F. Repeat steps D and E for all the organisms.

1. Which organisms in the web are the producers?

2. Which are the consumers?

3. Which are the decomposers?

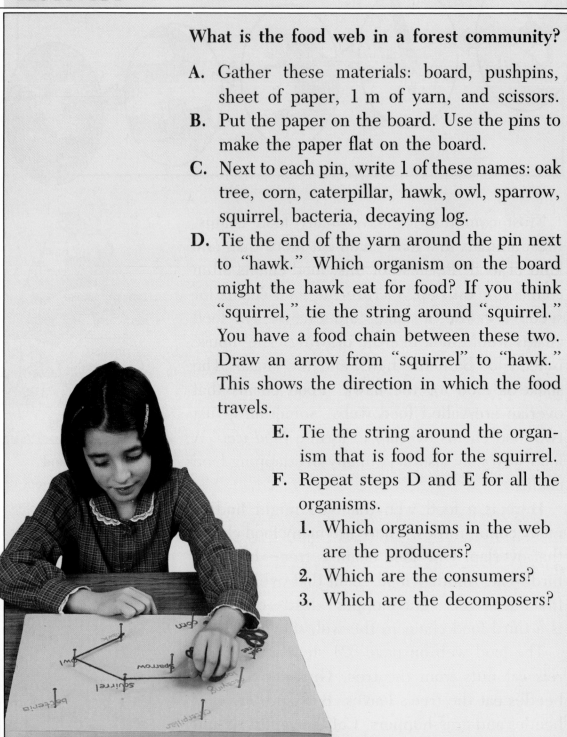

320

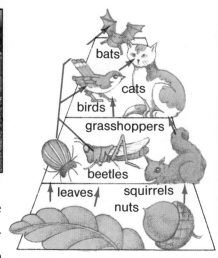

The leaves are the producers in four of the food chains shown here. The beetles, grasshoppers, birds, and bats each are members of two food chains. Cats are part of three chains. The nuts and squirrels belong to just one chain.

Food chains and food webs show us that life really depends on death. One animal eats another, and so on. An animal that kills another animal and then eats it is called a **predator** (**pred**-ah-tore). The animal that is eaten is called the **prey** (**pray**).

Predator: An animal that kills other animals for food.

Prey: An animal that is eaten by other animals.

Section Review

Main Ideas: Food chains that connect or overlap are called food webs. An animal is either a predator or the prey.

Questions: Answer in complete sentences.

1. Draw a food web for these organisms: hawk, snake, caterpillar, beetle, and seeds.
2. Which organisms in the above food web are predators? Which are prey?
3. What is the difference between a food chain and a food web?

19-3.

The Balance of Nature

Nature has ways of balancing itself. This farm cat is one way. She is a living mouse trap. She affects the number of mice that live in her community. What do you think will happen to the number of mice if the cat leaves or dies? When you finish this section, you should be able to:

☐ **A.** Explain the effect on a community food web if one part of it changes.

☐ **B.** Give examples of how nature balances itself.

Look at the food web shown here. It is a simple one. It involves water plants, algae, insects, frogs, snakes, and an eagle. The pond food web is the way all the organisms get energy to live. The eagle gets its energy by eating snakes. The snakes eat frogs. The frogs eat insects. The insects eat plants. What do you think would happen to the snake population if the eagles died? How would an increase in the snake population affect the frog population? Answers to these questions will help you understand how nature balances itself. Let's find out.

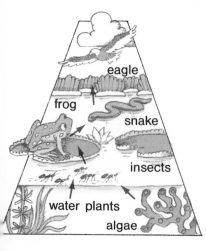

eagle
frog
snake
insects
water plants
algae

At one time, scientists believed that deer in Arizona were in danger of being killed off. Their predators were the puma, the wolf, and the coyote. Scientists tried to solve this problem by killing the predators. In less than 20 years, the deer population got much larger. They ate all the green plants around them. They even ate tiny, young trees. This destroyed the forest. The deer could not find enough food. They began to starve to death. Was killing the deer's predators a good idea?

This story shows that all plants and animals in a community depend upon each other to survive. If something happens to one member of a food chain or food web, other members are affected. Look what happened when the deer's predators disappeared from the food chain. The number of deer increased. But the number of green plants did not. The deer began to die because the forest was no longer able to feed so many of them.

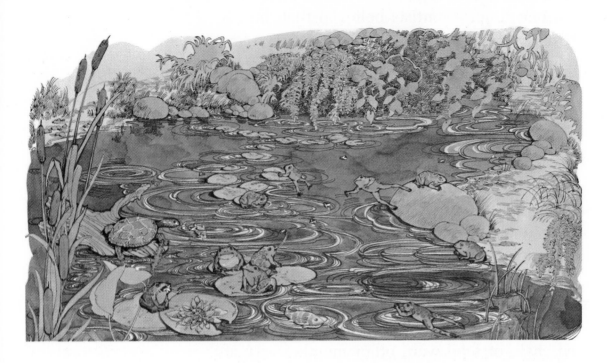

What do you think will happen if the number of frogs in a pond increases? The frogs will eat more insects. The number of insects will decrease. Then the frogs may begin to die. There will not be enough food for them.

The story of the deer shows that nature has a way of checking, or balancing, itself. The natural predators of the deer could have controlled the number of deer in the community. This is what is meant by the "balance of nature." Predators do not often kill more prey than they need to survive. There was no danger of the deer being killed off as the scientists thought. Without these natural enemies, however, the growing number of deer did not have enough food to survive. So nature stopped the rapid growth of deer by itself. Because of the lack of food, many deer died.

ACTIVITY

How does nature balance populations?

A. Gather these materials: graph paper and crayons.

B. Make a bar graph. Use the data shown in the chart.

1. During which year were many plants being eaten by the rabbits?
2. What happened to the plants during that year?
3. What effect did this have on the rabbit population?
4. How did nature balance the rabbit population?

A RABBIT POPULATION	
Year	Population Size (in thousands)
1920	20
1921	60
1922	65
1923	75
1924	60
1925	20

Section Review

Main Ideas: Animals and plants in a community depend upon each other to survive. If something happens to one member of a food chain or web, other members are affected.

Questions: Answer in complete sentences.

1. What would happen to the number of insects in a pond if the number of turtles increased?
2. What would happen to the number of green plants if the number of insects increased?
3. Draw a food web showing green plants, seeds, deer, and wolves. What would happen to the deer if the plants and seeds died?

People Affect Communities

As human beings, we are part of food webs and communities. We can affect communities like any other animals or plants. This lake was once clean. It had many fish and other forms of life. People could swim in the lake. Now look what has happened. How did people affect this lake community? When you finish this section, you should be able to:

☐ **A.** Identify the kinds of things *ecologists* do.

☐ **B.** Describe ways that human beings affect communities.

Over the past 20 years, people have become concerned about the environment. They realize that living things need a clean environment in order to live. Scientists who study living things and their environments are called **ecologists** (ee-**kol**-lo-jists). *Ecologists* study the air, water, and soil. They try to learn all they can about the needs of organisms. They know that human beings are also part of food chains.

The word *ecology* comes from the Greek word *oikos*. This word means "household, home, or place to live." So ecology deals with organisms and their environments. Ecologists study how living things within a population affect each other. They also study how different populations affect each other.

Ecologist: A scientist who studies living things and their environments.

All living things in the biosphere are connected. Living things also need non-living things like soil, water, and air. If something happens to soil, water, or air, living things will be affected. Let's see how humans can cause some of these changes.

Human beings make use of the environment in which they live. Unlike other animals, humans can change an area to meet their own needs. When they do this, it affects other living things in the community. Sometimes the effects are not good ones. The picture below shows how wastes from homes and factories may enter a river. What are some other ways a stream may become polluted?

Adding harmful things to air and water causes what is known as pollution. The water in polluted streams is unfit to drink. It also kills plant and animal life. Suppose the polluted stream water reaches a pond. What do you think would happen to its community members? What would happen to the food chains and food webs?

Air pollution is also a problem. Factories and cars give off harmful gases into the air. Insect poisons that farmers spray on their crops also pollute the air. Some of these poisons kill harmless animals as well as insects. They may also harm people.

ACTIVITY

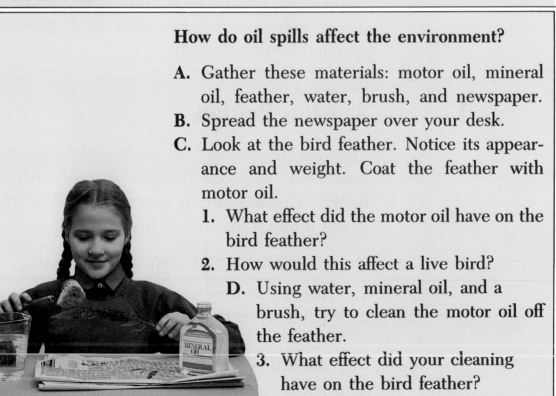

How do oil spills affect the environment?

A. Gather these materials: motor oil, mineral oil, feather, water, brush, and newspaper.

B. Spread the newspaper over your desk.

C. Look at the bird feather. Notice its appearance and weight. Coat the feather with motor oil.

1. What effect did the motor oil have on the bird feather?

2. How would this affect a live bird?

D. Using water, mineral oil, and a brush, try to clean the motor oil off the feather.

3. What effect did your cleaning have on the bird feather?

Everyone can help solve pollution problems. For example, control devices can reduce factory and car pollution. What can you do? Returning cans and bottles can help.

Section Review

Main Ideas: Human beings often affect the environments and communities of other living things. Humans are connected to all living things through the web of life.

Questions: Answer in complete sentences.

1. What is ecology?
2. What are two things that ecologists do?
3. What would happen to the food chains in a pond if the water became polluted?
4. What is an example of pollution?
5. How could wastes harm the environment?

CHAPTER REVIEW

Science Words: Unscramble the letters to find the correct terms. Match them with the definitions.

1. EDOOWBF (two words)
2. YREP
3. TERROPDA
4. CHIDOFANO (two words)
5. GELOCITSO

a. The path food travels in a community
b. Food chains that overlap
c. An animal that eats other animals
d. An animal that is eaten by other animals
e. A scientist who studies living things

Questions: Answer in complete sentences.

1. How is a food chain different from a food web?
2. Use arrows to show the food chain for a deer, a wolf, and leaves.
3. Draw a food web on a community chart for leaves, chipmunks, nuts, birds, worms, and a fox.
4. Which of the above organisms is a predator? Which are its prey?
5. Suppose the wolf population in a forest got very large. What would be the effect on the deer in the forest?
6. How would plants be affected by the change in the wolf population?
7. Would the population of wolves keep getting larger? Explain your answer.
8. How does pollution affect living things in a stream?

INVESTIGATING

How much do consumers eat?

A. Gather these materials: 2 crickets, wide-mouth jar with cap, mustard seeds, water, and cotton balls.

B. Make a cricket home. Put a wet cotton ball in the jar. Put about 40 mustard seeds in the jar.

C. Put 2 crickets in your cricket home. Cover the top of the home. Put holes in the cap so that air can get in the jar.

D. Each day for the next 5 days, count the number of seeds left in the cricket home. At the end of 5 days, put the crickets back in a terrarium, or keep feeding them in the cricket home.

    1. About how many seeds do the crickets eat each day?

    2. Do you think 40 seeds are enough food for 4 crickets for 5 days? Explain your answer.

    3. Frogs eat crickets. How many crickets do you think a frog needs to eat in a day to survive?

CAREERS

Forest Ranger ▶

Forest rangers are people who protect the plants and animals that live in forests. Some forest rangers help animals that are hurt or sick. Others take people on nature walks. They teach people about the kinds of living things around them.

Forest rangers also look for fires. Sometimes fires are started by careless people.

◀ Conservationist

Conservation means protecting or guarding. Conservationists try to protect endangered plants and animals. They work outdoors in fields, forests, or mountains. Some conservationists study how air pollution harms plants and animals. Others may go out in boats to test for water pollution. Conservationists also study how humans affect animals and plants.

GLOSSARY/INDEX

In this Glossary/Index, the definitions
are printed in *italics*.

Wedge (wej): *two inclined planes joined together to form a sharp edge*, 185

Wegener, Alfred, 25

Weight, measuring of, 176

Whales, 265, 275, 303

Wheel and axle: *a simple machine that is a wheel that turns on a rod*, 199–200

Wheels, and friction, 216–218

Wind, formation of, 133–135

Winding staircase, 183–184

Work, 175–176, 178–180

X rays, 63

Yeasts, 310

PHOTO CREDITS

Unit 4: p. 172–173 — HRW Photo by David Spindel; p. 174 — Karl Hentz/The Image Bank; p. 175 bottom right — Joe DiMaggio/Peter Arnold; p. 178 left — E. R. Degginger/Bruce Coleman, Inc.; p. 187 — John McGrail/Discover Magazine; p. 193 — Werner H. Muller/Peter Arnold; p. 206 — Richard Hutchings; p. 211 — Frank Siteman/Stock Boston; p. 219 — George Young/Colour Library International; p. 225 right — John Coletti/Stock Boston; left — Louis Fernandez.

Unit 5: p. 226–227 — Z. Leszczynski/Animals Animals; p. 228 — Kojo Tanaka/Animals Animals; p. 229 top — Breck Kent/Animals Animals; bottom right — Animals Animals; bottom left — Oxford Scientific Films/Animals Animals; p. 232 left — E. R. Degginger/Animals Animals; right — Manuel Rodriguez; p. 233 — Fritz Prenze Brace/Bruce Coleman, Inc.; p. 234 both — David C. Fritts/Animals Animals; p. 236 — Manuel Rodriguez; p. 237 top — D.P.I.; bottom — Harvey Lloyd/Peter Arnold, Inc.; p. 238 — Charles Palek/Animals Animals; p. 240 — Doug Hart; p. 242 — Ron & Valerie Taylor/Bruce Coleman, Inc.; p. 243 — Runk/Schoenberger/Grant Heilman; p. 244 all — Carolina Biological Supply; p. 245 left and right — Breck Kent/Animals Animals; center right — Ann Hagen Griffith/D.P.I.; center left — Edmond R. Taylor/D.P.I.; p. 247 — Breck Kent/Earth Scenes; p. 248 — Runk/Schoenberger/Grant Heilman; p. 249 right — Grant Heilman; p. 251 bottom — Edmund R. Taylor/D.P.I.; p. 252 — Carolins Kroeger/Animals Animals; p. 253 — Brian Milne/Animals Animals; p. 254 top — C. A. Morgan/Peter Arnold, Inc.; bottom — E. R. Degginger/Animals Animals; p. 257 — J. C. Stevenson/Earth Scenes; p. 260 left — Tom McHugh/Photo Researchers, Inc.; right — Leonard Lee Rue III/D.P.I.; p. 263 — C. Haagner/Bruce Coleman, Inc.; p. 264 left — David Overcast/Bruce Coleman; right — Zig Leszczynski/Animals Animals; p. 265 — Courtesy of the American Museum of Natural History; p. 266 top left — Runk/Schoenberger/Grant Heilman; center — Barry Runk/Grant Heilman; bottom — Inger McCabe; p. 269 top right — Manuel Rodriguez; bottom left — John S. Flanner/Bruce Coleman, Inc.; bottom right — Phil Dotson/D.P.I.; p. 271 — M.W. F. Tweedie/Bruce Coleman, Inc.; p. 272 — Jane Burton/Bruce Coleman; p. 273 left — Bruce Coleman, Inc.; right — Runk/Schoenberger/Grant Heilman; p. 274 left — Jeff Foote/Bruce Coleman, Inc.; right — Clyde H. Smith/Peter Arnold, Inc.; p. 275 — Ray Gilbert/Photo Researchers, Inc.; p. 279 right — Richard Choy/Peter Arnold, Inc.; left — Manuel Rodriquez.

Unit 6: p. 280–281 — T. N. Liversedge/Photo Researchers; p. 282 — E. R. Degginger/Bruce Coleman, Inc.; p. 283 left — Runk/Schoenberger/Grant Heilman; right — Phil Dotson/D.P.I.; p. 285 top — William E. Ferguson; p. 286 — H. Pleischinger/Peter Arnold, Inc.; p. 288 — Manfred Kage/Peter Arnold, Inc.; p. 291 — Bettmann Archives; p. 292 — Jim Zipp/Photo Researchers, Inc.; p. 293 left — Peter Arnold, Inc.; p. 294 left — E. R. Degginger/Animals Animals; right — Timothy Egan/Woodfin Camp & Assoc.; p. 298 — Lowell Georgia/Photo Researchers, Inc.; p. 303 top — Animals Animals; bottom — R. Ellis/Photo Researchers, Inc.; p. 304 bottom — Runk/Schoenberger/Grant Heilman; p. 305 — Manfred Kage/Peter Arnold, Inc.; p. 310 all — Runk/Schoenberger/Grant Heilman; p. 314 — W. E. Harvey/Photo Researchers, Inc.; p. 321 — Grant Heilman; p. 323 — Paul Dotson/D.P.I.; p. 326 — Dan Budnik/Woodfin Camp & Assoc.; p. 328 — James Karales/Peter Arnold, Inc.; p. 329 — USDA Photo; p. 330 — Environmental Action Coalition.

ART CREDITS

Anthony Accardo, pages 72, 76, 80, 83, 88, 93, 104, 135, 188, 208, 212, 216, 270, 277, 316, 332

Gary Allen, pages 19, 25, 26, 28, 29, 32, 51, 81, 84, 85, 140, 145, 150, 197, 210, 217, 218, 258

Ebet Dudley, pages 7, 12, 22, 29, 32, 34, 35, 37, 73, 77, 87, 90, 102, 116, 147, 148, 158, 182, 192, 196, 204, 209, 230, 238, 245, 247, 250, 255, 259, 278, 286, 294, 300, 301, 305, 307, 315, 317, 319

Morissa Lipstein, page 32

Jan Pyk, pages 308, 316, 318, 319, 321, 322

Pat Stewart, pages 13, 71, 75, 82, 94, 98, 118, 123, 130, 131, 139, 140, 141

Vantage, pages 4, 5, 7, 79, 85, 89, 90, 91, 95, 96, 105, 108, 109, 115, 124, 154, 187, 189, 191